Visions of a Rainforest

Visions of a Rainforest

A Year in Australia's Tropical Rainforest

Text by
STANLEY BREEDEN
Illustrations by
WILLIAM T. COOPER
Foreword by
SIR DAVID ATTENBOROUGH

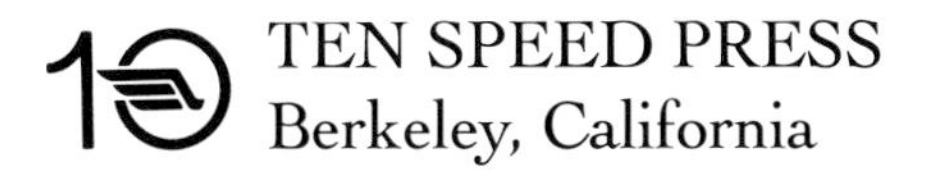
TEN SPEED PRESS
Berkeley, California

HALF-TITLE PAGE ILLUSTRATION:
The red-eyed tree frog sings from leaves hanging over the pond.

TITLE PAGE ILLUSTRATION:
The emergent fig tree with the slopes of Mount Bartle Frere in the background encapsulates tropical rainforest. The early morning sun disperses the last mist from the valleys. The fig is strangling its host tree and has grown to be among the largest in the forest. Epiphytic ferns and orchids cling to its limbs and trunk. Lianes, including the pink flowering jungle vine, have climbed their way up. In the centre foreground are the feathery leaves of a climbing palm, the wait-a-while. In the lower right is a tree with pink new shoots. Emergents, stranglers, epiphytes, lianes, colourful new growth are all characteristics of tropical rainforest. They define it.

VISIONS OF A RAINFOREST
First published in Australasia in 1992 by
Simon & Schuster Australia
20 Barcoo Street, East Roseville NSW 2069

Text and cover design by Steven Dunbar
Map by Dianne Bradley
Typesetting by Everysize Typeart Services, Sydney
Produced by Mandarin Offset, Hong Kong

Ten Speed Press
P.O. Box 7123
Berkeley, California 94707

FIRST TEN SPEED PRESS PRINTING 1993

Library of Congress Cataloguing in Publication Data
Available on request

ISBN 0-89815-523-1

Printed in Hong Kong

1 2 3 4 5 – 96 95 94 93 92

Contents

Topknot pigeon. Flocks of these birds come to Bulurru's fruiting trees.

Foreword

It seems extraordinary that one small patch of bush, up in northern Queensland, should so recently have become of compelling interest to naturalists around the world. It amounts to barely more than 7500 km^2, a minuscule area compared with the 700 000 km^2 of forest that, just to the north, blanket the island of New Guinea. Indeed, until a few years ago, the Australian patch was often dismissed as being little more than an overflow, somewhat depleted of species, of the huge New Guinea jungle. It was not until we all began to realise how rich rainforests are and became alarmed at how rapidly they are being destroyed that naturalists started to look really hard at that which grew in Australia.

When they did, they were astonished. The rainforest had very little to do with New Guinea. Over a thousand different species of trees grow in it and about half of them are found only in Australia. Not only that, but a great number of its animals and plants live nowhere else at all. This fragment is all that remains of a much greater forest that once spread over a large part of the continent but which, during the last 140 million years, slowly contracted as Australia moved gradually northwards and became drier and drier.

As the special character of this forest was more widely recognised, a few discerning naturalists went to live within it. Stanley Breeden was among them. Expert and knowledgeable as he was about Australia's natural history, he found initially this thick wet tangle of trees a very baffling place. It is an experience that most of us feel when entering a rainforest for the first time. Indeed, only too many of us remain in that state of bafflement. But Stan, by patient, systematic and meticulous observations, day after day, learned to recognise species after species and slowly built up a coherent picture of this complex jigsaw of animals and plants.

One of his neighbours, William Cooper, joined him in the long process of discovery and recognition. Bill's superb illustrations of birds have earned him an international reputation. Those of us who treasure the monumental books containing his portraits of all the world's parrots, kingfishers and birds of paradise, have also come to take particular delight in the tiny extra details that he occasionally tucks away in an unobtrusive corner of a picture—the delicate filigree pattern of a lichen encrusting a twig, a strangely-shaped fruit, a spider spinning a web or a caterpillar industriously nibbling a leaf, unaware of the huge beak of a bird just above that is about to make a meal of it. For us, the pages that follow will bring particular delight for here, at last, that eye for detail is allowed full scope. The results are enchanting.

So Stan's pen and Bill's brush between them have produced a marvellously vivid picture of what it is like, not just to visit a tropical rainforest, or even to live within it, but to explore it, to sense its cycles, to watch its growth, to witness its dramas. It is more fascinating even than that, for this a most particular and special rainforest. It is the one from which many of the animals and plants that populate the rest of Australia are ultimately derived. This, you could argue, is a portrait of Australia's Garden of Eden. Long may it remain undespoiled.

David Attenborough

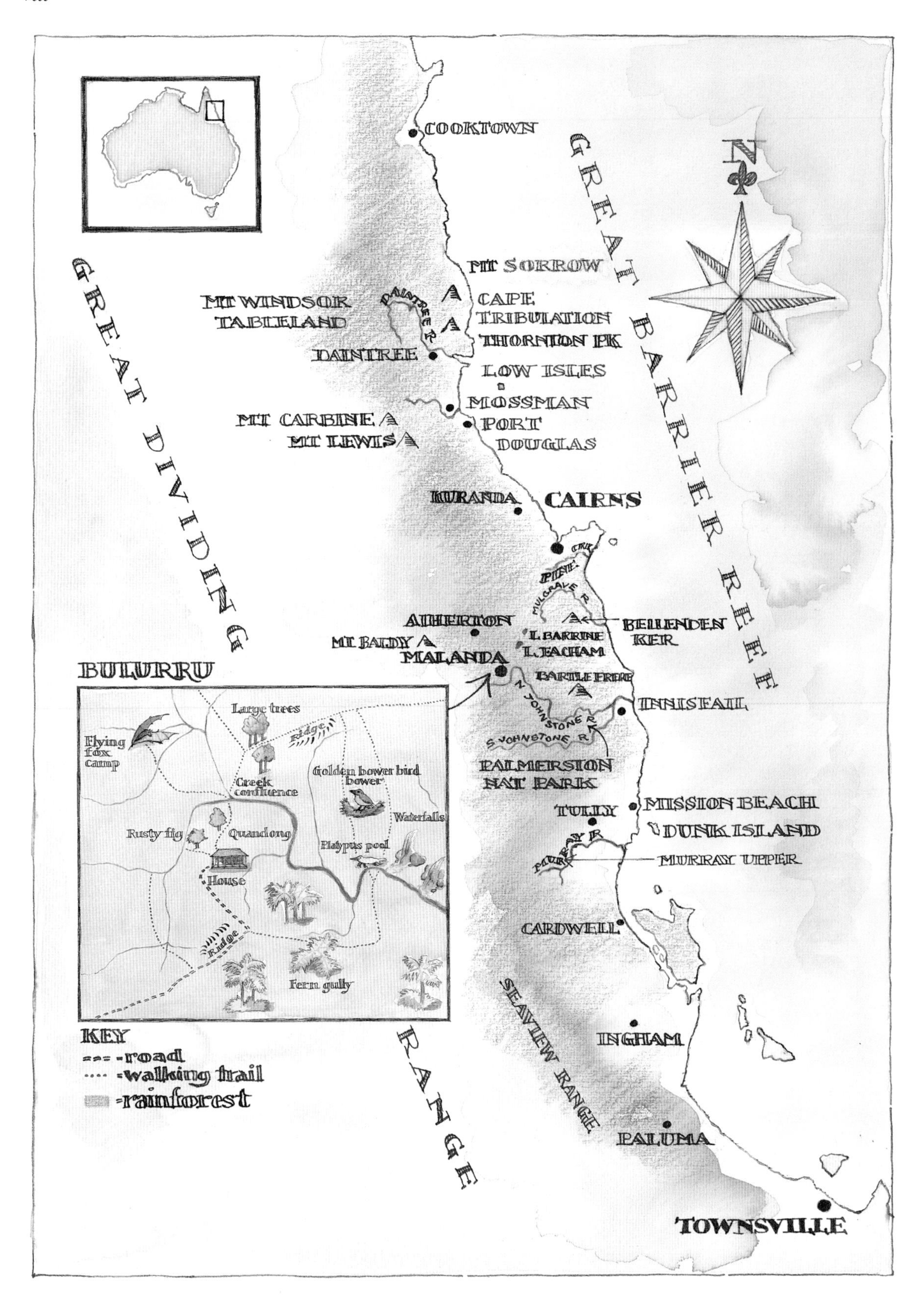
COOKTOWN
GREAT BARRIER REEF
N
MT SORROW
MT WINDSOR TABLELAND
CAPE TRIBULATION
THORNTON PK
DAINTREE R
DAINTREE
LOW ISLES
MOSSMAN
MT CARBINE
MT LEWIS
PORT DOUGLAS
KURANDA
CAIRNS
GREAT DIVIDING RANGE
MULGRAVE R
ATHERTON
BELLENDEN KER
MT BALDY
L BARRINE
L EACHAM
MALANDA
BARTLE FRERE
INNISFAIL
N JOHNSTONE R
S JOHNSTONE R
PALMERSTON NAT PARK
TULLY
MISSION BEACH
DUNK ISLAND
MURRAY R
MURRAY UPPER
CARDWELL
SEAVIEW RANGE
INGHAM
PALUMA
TOWNSVILLE
BULURRU
Large trees
Flying fox camp
Ridge
Creek confluence
Golden bower bird bower
Waterfalls
Rusty fig
Quandong
Platypus pool
House
Ridge
Fern gully
KEY
=road
=walking trail
=rainforest

Preface

Tropical rainforest has more species of plants and animals than any other land habitat, but at the same time less is known about them than those of more familiar places such as eucalypt forest and heathland. Because of this, many rainforest species do not have common or English names. In some cases, especially among insects and plants, they do not even have scientific names. This poses a dilemma for the natural history writer: the use of common names can be inaccurate and limiting, but scientific names can be disruptive to the flow of the text. I have made a conscious effort to use common names wherever possible and to use scientific names only where no common names exist. A list of scientific names for the species of plants mentioned is given at the end of this volume.

I am grateful to the people who gave so willingly of their time, their knowledge and their thoughts in conversations with me. Their thoughtfulness broadened my appreciation of the rainforest and its workings. Geoff Monteith and Geoff Tracey were especially generous in their assistance. I am greatly indebted to Belinda Wright for helping to establish Bulurru and for her unstinting work in designing and setting up what has become my ideal home in the rainforest. I much appreciate Robyn Russell's help in bringing the text under control.

It has been a long-held wish to collaborate with Bill Cooper on a project that is close to both our hearts. The perfect opportunity arose when we came to live in tropical rainforest, and not accidentally, as neighbours. I am also indebted to Bill and his wife Wendy for their invaluable support throughout the writing of this book, for contributing many observations and for the fruitful discussions in trying to unravel the mysteries of the rainforest.

Stanley Breeden
Bulurru, April 1992

Ulysses butterfly with corkwood leaves which are the food of its caterpillars.

Finding the Right Place

Far north Queensland's rainforest has been one of my favourite places ever since I first went there in 1958. At that time I travelled north from Brisbane to photograph and to collect natural history specimens for the Queensland Museum. Working mostly with dead animals and being part of the public service bureaucracy finally palled. I resigned in 1968. Within months I was back in the tropical rainforest where I spent most of the next two years gathering material for *Tropical Queensland*, a book I wrote about the area's wildlife. My work of photographing, filming and writing about natural history then took me to other parts of the world, but I vowed that one day I would go back.

The opportunity did not come until late in 1987. I then decided that after almost twenty years of itinerant but rewarding life in Australia, India and the USA, it was time to set up a permanent base. I needed a place in the bush where I could photograph and write about wildlife at my doorstep, but one that was not so remote that I could not communicate with the rest of the world.

Because there is a greater variety of plants and animals in the tropical rainforest than anywhere else, it was the logical place to go. The statistics are impressive. In Australia, tropical rainforest is found only between Townsville and Cooktown and takes up just one thousandth of the land area. Yet twice as many species of frogs live here than in the whole of the state of Victoria. There are more species of butterflies in these forests than in any other habitat. One-third of the country's marsupials, one-fifth of the birds, one-quarter of the reptiles, and two-thirds of the bats make their home here. Two-fifths of Australia's plants occur in this small corner. The trees alone are so diverse that just one plot the size of a generous suburban house and garden may contain 165 different kinds. A tall eucalypt forest of almost equal density, by contrast, is often dominated by just three or four species.

So when I decided to settle in north Queensland, I had visions of living deep in the forest surrounded by tall trees. There would be butterflies and brilliantly coloured beetles swarming along the forest edge, orchids smothering tree branches, exotic possums peering down from vines, huge pythons slithering through the undergrowth, tree-kangaroos leaping through the canopy, cassowaries looking into my windows, frogs singing on wet nights; in other words, it would be a naturalist's paradise.

The search began for this ideal place. The criteria were strict. It had to be on the Atherton Tableland. The coastal lowlands are too hot, birds and possums too few, and there are no golden bowerbirds at all. Mosquitoes and green tree ants, so prevalent and annoying along the coast, are absent at these higher altitudes, but they do have their fair share of leeches, ticks and those supremely irritating little mites known as scrub itch. On the other hand the place should not be too high above sea level, for then it would be cold and constantly misty; there would be fewer butterflies and other interesting insects, and the forests would be less diverse. The place should be at the end of a road, well away from passing traffic, and adjoin the World Heritage rainforest.

I looked from one end of the Tableland to the other. I saw parcels of rainforest on high mountain ridges and in lower altitudes. There were ones with high rainfall and others with barely enough to maintain rainforest. None was exactly what I needed. In the end I returned to a place I had been shown by friends on my first day. It had looked unpromising but I thought maybe if I explored a bit further it could possibly be the right place. At its point of access, at the end of a grass track, there was an impenetrable tangle of lantana and other virulent weeds. This thicket occupied about six of the block's 60 hectares and it was difficult to get beyond it into what should be rainforest.

One day two friends and I went to have a better look. Our first objective was to track the source of the sound of rushing water. Was it a waterfall, a series of rapids? After a struggle down a steep slope through thick undergrowth we discovered it was both—and much, much more. There was not one waterfall, but two, as well as rapids and large clear pools surrounded by giant trees. It was one of the most beautiful and romantic places I had ever seen. The stream and the forest on its opposite side, still on the same piece of land, looked wonderfully promising. Beyond the waterfall was World Heritage forest in the shadow of Queensland's tallest peak, Mount Bartle Frere. The only drawbacks were that the forest had been logged about 15 years ago and that it was very, very wet indeed; on average more than 4000 millimetres of rain a year. But that, and the comparatively low altitude, meant that this was the most complex rainforest on the Tableland. Another worry was that at the point of access the place was a mess. The lantana was so dense that it was impossible to get an idea of the lie of the land. Was it steep or gently sloping? Was there a site I could build a house with a view of the forest? There was no way of knowing. I took that part on faith and bought the place. I called it Bulurru, meaning 'protective benevolent spirit' in a local Aboriginal language.

Lantana proved ridiculously easy to push over. The difficult part, at first, was to persuade the bulldozer driver to leave the trees, particularly the small ones, standing. He said it would be much quicker to just push the whole lot down. But he was a kindly and sympathetic man and he soon understood what I was trying to do.

At the end of the day all lay revealed: gentle slopes of rich, red-brown, volcanic earth around islands of trees. I still did not know the name of a single one of those trees, but in the years to come they would provide me with great spectacle and immense pleasure.

Standing on top of a ridge, I selected a place to build my house. It was surrounded by groves of trees that attract all kinds of animal life. There was a commanding view of the rainforest and in the distance I could hear the whisper of the waterfalls and rapids. I had to imagine green grass growing where so far there was only brown soil. It was one of those rare moments when you suddenly realise that a long-held dream will almost certainly come true. These feelings were even more poignant than when I finally saw the house built and surrounded by green grass. I suppose that was because the building process was slow and the changes were gradual. On the day of the bulldozer the seemingly miraculous change was as fast as it was unexpected.

Spectacled flying-foxes feeding on the flowers of a blue quandong. There is a large flying-fox colony at Bulurru.

The building and lawn-growing took about two years. During that time I cleared old logging tracks and made narrow trails to favoured spots. On a ridge close to the waterfall I cut down a particularly rampant patch of lantana and discovered breathtaking views of a gully, full of a surprising variety of ferns and orchids. At a far corner I found a flying-fox colony and spent many days watching these intelligent animals. A grey satinash of truly gigantic proportions that had eluded the loggers remained in a ravine. The perennial stream that runs the full length of Bulurru was a constant, ever-changing delight.

During my early explorations Bulurru seemed limitless. I walked for days without seeing the same places. Wandering along the ridges, down steepsided gullies or along winding creeks there was something new to see around every corner, and in this rugged terrain the corners were not far apart. It might be an imposing tree, some colourful fruit, a rock formation, a pool, an insect, a flock of birds, a small mammal scurrying among the leaf litter, that made me gasp with wonder. A peculiarity of a place of changeable topography and dense vegetation, I discovered, is that it seemed much larger than it actually was. The main reason for this was that I could not see very far ahead; I could not get an impression of the place as a whole. In open grassland, on the other hand, it would be easy to see from one end of a 60-hectare patch of land to the other, and beyond.

The islands of trees, no longer suppressed and strangled by the lantana, put on new leaf and spread their limbs. Some islands were only a single tree, a celerywood or a tree fern, while others were miniature forests a quarter hectare in size.

One of the pleasures of gardening is to establish shrubs, trees, or whatever plants that appeal to you most, and then watch them grow to maturity. My gardening was rather unconventional and to me its pleasures were and remain even greater. I did not plant anything. My pleasure was to see what plants came up in the various islands, how these flourished and spread and what flowers and fruits each produced and which animals these attracted. There were wonderful surprises. A rare ground orchid sprang up right beside the house. Gingers and ferns, palm-lilies and red-fruited sedges, native lasiandras and zamias have established a shrub layer. Moss clothes old logs and stumps in green velvet.

The resultant garden is an amazing and fortuitous mixture of flowering and fruiting plants that attract a brilliant array of animals, mostly fruit and nectar eating birds. Friends often comment that even the most careful and thoughtful planning could not have improved on the design and species composition of the 'garden'. As it happened, it was a gift from the natural forces that shape the rainforest. Even in gardening it is sometimes difficult to improve on nature.

Over those two years, I also observed the seasons. As people of mostly European descent we tend to think of the seasons in the classical pattern of a temperate climate. But this is not so clear cut in the tropics and we should invent new words to describe the seasons here. We could just say that there is a wet season and a drier one. But it is more subtle than that. There are three wet seasons and two drier ones. The first wet season begins in mid-November and brings a succession of often violent thunderstorms. The days are hot, the hottest of the year. When the wind is from the west, temperatures can reach 37°C. This is followed by the true monsoon when blankets of clouds roll in from the north and northwest and bring days and days of heavy rain. Creeks and rivers flood. The monsoon can start any time from the end of December, but usually does not arrive till early February. It departs again in early April. February and March are our months of heaviest rainfall. Last February Bulurru received 1050 millimetres of rain.

In April our third wet season begins. From then until early June we can expect weeks on end of mist and light drizzly rain. But it is a warm rain, especially early in the season. Bulurru and its surroundings are especially susceptible to this drizzle. Through a quirk of geography, Mount Bartle

Frere and the constant southeasterlies blowing in from the Pacific Ocean conspire to channel these rains in our direction. Just a few kilometres away not even a third as much rain falls. In conventional, temperate climate terms April is in autumn; here it is a time when many of the trees and vines put on new leaf. Quite a few are in flower.

Superimposed on these wet seasons is a fourth one of occasional but catastrophic violence. Tropical cyclones can spiral in from the ocean any time between December and May. They are completely capricious and unpredictable. Most years at least one cyclone strikes the lowlands, but their destructive force rarely reaches the Tableland. The last one to hit Bulurru was cyclone Winifred in February 1986.

Even though it may rain one or even two weeks at a time, there are always sunny breaks. March and April are our months with most rainy days, but even then, on average, these months have eight rainless days each. These are the days I rush out and revel in the forest's freshness, in the new growth and the animals' seeming joy at the relief from wet weather. But even on days of drizzle, sheltering under an umbrella and wearing rubber boots to keep the leeches at bay, I enjoy and appreciate the special character and feel of the forest. It is when the forest seems at its most dynamic, when it is truly *rain*forest.

Fruits and leaves of the white supplejack. These vines are common in the islands of trees at Bulurru.

The drier seasons—there is no completely dry one—are comparatively brief, lasting from about mid-June to mid-November. During June, July and early August there are lengthy periods when the skies are absolutely clear and there is very little wind. Mount Bartle Frere then seems etched against the sky. The days are balmy. The night sky fairly crackles with stars. Early mornings are crisp and on one or two days frost dresses the grass in low-lying areas in white. Frost, however, never invades the forest. The winter's coolness is magnified by the Tableland's higher elevation; the lowlands remain pleasantly mild.

In September the weather slowly warms. Bird song swells in volume and variety. Nesting begins and continues into January for most birds. You might think this warm, dry season is spring, but then this is the season when there is the greatest leaf fall. The trees, of course, are evergreen and do not shed all their leaves. But enough do fall to make the forest noticeably lighter.

However, we are far enough from the equator to have faint traces of the temperate climate pattern. We are in the tropics, but not in the equatorial zone. Days are longer in summer than in winter. The winter sun is lower in the sky, which means that steep slopes facing the south and deep gullies receive little or no sunshine for several months. So even though our thinking is mostly of wet times and dry times, we also talk of summer and winter, spring and autumn.

Finally, on the tenth of December 1990, two and a half years after I bought Bulurru, I moved into my house. It has become the naturalist's paradise I had hoped it would be. When I wake in the morning I watch purple-crowned pigeons feeding in bollywood trees only a few metres from my bedroom windows. I go to the kitchen and see pademelons, small rainforest wallabies, grazing at the forest edge while I have my breakfast. I go to my desk and find a pair of bright yellow sunbirds hovering at the window. This corner of the house is shaded by trees. Among them are several with heavy crops of fruit which bring all kinds of birds to within metres of where I sit and write these pages. Across a patch of lawn stands a very special tree, a large rusty fig. It gets its name from the rusty-brown colour of the undersides of its leaves.

My good friends of many years, Bill and Wendy Cooper, live about two kilometres away across the ridges. Their place, which they have called Chowchilla, after a local bird, is also rainforest. Bill, one of the world's leading wildlife artists, has illustrated this volume with his masterful and authentic drawings. Wendy has become adept at the difficult task of identifying rainforest plants. She and Bill have taught me just about everything I know about the identities of the trees, vines and shrubs.

Although I have wandered extensively over Bulurru for more than two years and have come to love all I discovered, I was so overwhelmed by it all that my observations have been rather disorganised. The result is that the place is still somewhat of an enigma. Today it feels like a brand new shoe, shining but still to be worn in. I sense that in time these forests, as I am drawn deeper and deeper into them, will be as comfortable as an old shoe. I also realise that to forge such an intimate relationship will require a conscious effort; knowledge and understanding have to be searched for.

There are a number of ways you can gain insight into a subject and a place, to dig beneath its surface features. This is different for different people. For an artist, like Bill Cooper, it is to draw. He once told me that drawing and painting in a naturalistic style, as he does, you must look more closely at everything than most people do. His first impulse on seeing an animal or plant of particular interest is to draw it as accurately as he can. In doing so he is aware of his subject's every detail, much more so than someone who does not draw. Whatever you draw, Bill said, you see accurately and in this way you really get to know the details of the life around you. His close observations, particularly in the tropical rainforest, have given Bill a greater understanding of the place and certain feelings of affinity with it.

I cannot draw. But I like to write and for the kind of writing I do, I also have to observe closely. It is to come to terms, in my own way, with the tropical rainforest that I begin this diary. Writing about what I see and what I learn will, I hope, sort it out in my mind. The first entry is made two days after I move in.

To place both Bulurru and my own thoughts in a wider context, I visit other places and talk to people who, like me, have a passion for rainforest. Each place and each person brings a different perspective. The people I have conversations with tell me about aspects of the forest and express ideas that I could never have discovered on my own. I am also fascinated by why and from what background people come to live and work in the rainforest. Through Bulurru, my visits to other places, and conversations with friends, I hope to discover the tropical rainforest, the most complex of all habitats.

Into the Rainforest

BULURRU, 12 DECEMBER

When I awoke this morning after only my second night spent at Bulurru, I was filled with exhilaration and a feeling of expectation. As the rising sun bursts from behind the purple eminence of Mount Bartle Frere a light mist still swirls in the valley. Only the tops of the tall tree ferns and the elegant celerywoods rise above it. Rays of light slant through the foliage of the rusty fig. The dew is heavy and drips from the trees. Sweet scents of cunjevoi flowers drift in from the forest. It feels and looks as it does after the most refreshing of showers. But there was no overnight rain. The sky is cloudless and later it promises to be hot.

A cunjevoi from the forest understorey. Its flowers are sweetly scented.

Leaving the house, I nearly trip over a flock of red-browed finches. Some hop mouse-like over the verandah floor picking up tiny particles of grit. The thirty or so birds fly only a few metres away, just far enough so I will not actually step on them. Crossing the lawn I pause to watch a red-legged pademelon fossicking under a fruiting Boonjie tamarind. He finds one of the orange-red fruits that must have fallen during the night. A white-headed pigeon searches the grass for fallen celerywood seeds in an ungainly waddle. A pale yellow robin comes over and clinging sideways to a tree trunk has a close look at me. Chowchillas interrupt their scratchings among the leaf litter to engage in a loud, wild chorus of song. Flying low along the path a wompoo pigeon comes straight towards me, seemingly aiming for my waist. At the last minute it veers to one side and shows off the bright yellow under its wings and its purple chest. In a pink ash a brown pigeon calls to his mate. It sounds as if he says 'Mr Wolf, Mr Wolf, Mr Wolf'. She is not very responsive and the two chase each other from branch to branch with much noisy wing slapping. I silently salute the resident amethystine python who lies tightly coiled beneath a native lasiandra bush.

I decide on impulse to take a long exploratory walk over Bulurru. But it is with a greater sense of purpose than in the past. Until now I had wandered about rather aimlessly; there was so much to see, so much going on that I was confused as well as enthralled. Today I will take stock of the place, survey it and try to find some cohesion in what I see.

Leaving the open grassy areas behind I enter the forest. Overhead the crowns of six or seven tree ferns, about ten metres tall, meet in a lacework of green dappled by the sun. A scrubfowl strides in its high-stepping, head-bobbing walk across the track. Moments later his slow cackles come from the undergrowth. A palm-lily with bunches of intense red fruit, so shiny you can almost see your face in them, glows among the dark trunks.

At the confluence of the creeks I enter what is very close to primary forest. Only a few trees were felled here about 15 years ago. It is cooler, darker here. The trees are enormously tall, their canopies meeting 25 or 30 metres above me. The orchids, bird's nest and basket ferns growing on the branches of a buff beech leaning over the creek form enormous clumps. Some of the ferns may be a tonne in weight when saturated on a wet day.

The rapids where the two creeks meet and the large circular pond below them are in deep shade. Often platypus swim and dive here, but not this morning. A rifle bird rasps his strange call. Three species of quandong, a pink tamarind, three kinds of silky oak, a poison walnut, a mountain mangosteen, a silkwood, and many other large trees surround the confluence. Three enormous orania palms, a species that thrives only in the damp and darkness of a closed forest, stand like sentinels beside the stream.

Climbing pandans hang from tall trees. The sun has risen enough for shafts of light to pierce the dark forest. One such spotlight illuminates the growing tips of another vine, an Austral sarsaparilla, which has worked its way through the pandans. The new leaves are a vibrant red and look like drops of blood on the green curtain.

I cross the creeks by jumping from rock to rock. On the opposite bank the trail meanders steeply uphill—wild gardenias, mountain ardisias, wilkieas, zamias and small saplings struggle for existence among the straight trunks of the trees. At this point I always feel like I am entering another world—a mostly benevolent world full of mystery where new and exciting things wait to be discovered.

The top of the ridge is actually a high knoll, the highest point on Bulurru. Here grows one of the largest and most beautiful trees, which I have not yet been able to identify. It must be ancient. For about 20 metres it rises in a solid, symmetrical column before the first branch appears. Projecting as it does above the general canopy on this exposed part of the ridge, its crown has been battered by many a cyclone and storm. Around its base lies a tangle of dead branches that probably came down in the most recent storm. For such a giant tree, it has few branches and these are topped with small tufts of leaves. The pale red-brown bark is smooth without being shiny as a eucalypt's might be. Wherever it is free of the patches, medallions and frills of mosses and lichens, it has a soft patina. Halfway up the trunk is a large hollow, its entrance well worn by the comings and goings of some animal. Is it the home of a possum, or perhaps a sulphur-crested cockatoo? Standing between the sturdy but narrow buttresses at the base of the tree I feel dwarfed, and somewhat in awe not just of the tree's size but also its age. Is it 200 years old, 500? It could even be a thousand.

The trail dips gently then levels out into a broad ridge. Sheltered by the knoll, the trees have grown straight with wide crowns. These may have felt strong winds, but not the full wrenching force of a cyclone.

Two of the trees, standing just off the trail, are huge Kuranda quandongs. Large as they are, they are screened by the undergrowth, and are not discernible unless right beside them. They are outstanding for the enormous bulk of the dark brown boles at the bases of their trunks, which are covered in knobs and bumps. Fine rootlets reach down to the soil while new coppice shoots strain for the light above. Some of these boles, and the plank buttresses of other species, have the solidity and seeming permanence of stone. As they loom out of the forest understorey, they seem to be wooden rocks.

It is these solid columns and, in truly undisturbed forest, the spaciousness that make people reach for the simile that likens the forest to a cathedral. But is it really? To me not at all. Certainly not in structure; perhaps for some in the spiritual feelings it evokes.

A pair of chowchillas scratching in the leaf litter look for insects and other small animal life. The female has the orange chest.

Rain at the creek above Bulurru's waterfall.

The sun shining through red and yellow-green new leaves may remind me fleetingly of stained glass windows. Should we enter the forest in hushed reverence as we might a cathedral? Certainly not. Enter it non-aggressively and with a free and questioning spirit, enter it as a living vibrant place full of life, colour, drama and excitement. Rainforest is not built, it grows. It is not on a human scale, it is beyond that. It dwarfs but does not intimidate. The springy carpet underfoot is made up of dead leaves and twigs yet is alive with all manner of growing organisms. Often it is strewn with brightly coloured fruits that look like jewels. Birds sing and call exuberantly. Trees large and small are interwoven with knotted, twisted slings of vines. Everything is encrusted with epiphytes. The sweet scent of flowers wafts in the air, as does the slightly musty smell of decay. Rainforest is alive, vibrant, wild and untamed.

Bird's nest ferns growing on a small tree.

Just a few metres from the quandongs stands another large tree, one of the largest on Bulurru. Again I do not know its name. It is perhaps my favourite. The bark, like that of so many trees here, is a warm buff-brown in colour and smooth except in places where it is cracked and fissured like old skin. The strangest thing, however, is that although it grows as a single tree, it is not a solid straight column; it looks as if a series of thick, pliable stems have been twisted and braided together and have then become fused. As these 'stems' twine around each other, they go up and up and up, thirty metres, giving the trunk and main branches a fluted appearance. That is how it looks, but the trunk is a single solid structure.

Reaching down its trunk, as though anchoring the tree, are the almost white guy ropes of a strangler fig. It has taken root high up and is gradually wrapping the trunk in its inexorably multiplying roots. Many years from now the tree will have been strangled to death, its place usurped by the fig.

I make my way back to the trail. It is a slow process as I am waylaid several times by the thin, whiplike, hook-studded appendages of wait-a-while vines.These climbing palms are aptly named as every time you become entangled, it takes a while to unhook yourself.

The next thing to catch my eyes is a female birdwing butterfly fluttering among the shrubs at about shoulder height. She hovers around a large heart-shaped leaf, briefly settles and deposits an egg on its underside. The egg is pale yellow and about the size of a matchhead. So absorbed is she in her activity that I can approach to within a metre. Being so close to the largest of all Australia's butterflies never fails to make my heart beat a little faster. Having laid her egg the butterfly flies higher, following the vine. She finds more leaves and lays more eggs. The vine is a native Dutchman's pipe, of which several kinds grow here.

Following the flight of the butterfly, my attention is drawn to bunches of small red fruits just below the canopy. I walk into the forest and soon find some of the tiny glossy berries that have fallen to the ground. When I crush one with my fingernail it gives off a pleasant peppery scent, not at all pungent. The fruits are in fact from a native pepper vine. The black pepper we like to spice our food with is from a vine of the same genus that scientists call *Piper*. Worldwide there are some 2000 species, but only nine occur in Australia. All of them are mild in flavour. Some of the Australian peppers grow to be among the largest vines, their thick, woody yet pliable stems reaching lengths of 70 metres as they climb up the trees to reach the light. Once they reach the canopy their rampant growth may smother the crowns of two or three large trees.

The ridge continues to slope gently down before levelling out once again. There is a flat area here ideally suited for tree growth. Sixteen large straight trunks crowd together in an area the size of a tennis court: blush alder, tulip oak, tulip kurrajong, grey carabeen, blush silky oak, paperbark satinash, brown walnut, Queensland maple, nutmeg and others. The trees, so different in leaf, fruit and flowers, have trunks that are remarkably similar in appearance. All are brownish, straight, slender columns, all are more or less smooth, though one or two have small whitish pimples arranged in rows up and down or across their bark. As always, much detail is hidden by the scribbles of lichens and mosses. But there is an exception to this general pattern. The paperbark satinash has orange-brown bark which stands out vividly. This bark flakes and peels off in tissue-thin layers, giving the tree a shaggy appearance and allowing no foothold for any lichens or other epiphytes.

It is now warm, but not yet hot. Sun dapples shine through the leaves, making the whole forest a luminous green. Just as I am thinking of looking for a log to sit on to enjoy this peacefulness, there is a muffled drum roll. It seems quite close. There it is again, 'boom, boom, boom, boom', right behind me. I spin around and look straight into the light brown eyes of an inquisitive cassowary. The bird is nearly as tall as I am and of a heavier build. How silently it approached! We look at each other for a while. Lowering its head with neck outstretched almost to the horizontal, it booms again. It is an eerie sound, especially when, as happens most often, you cannot see the bird. The sound carries far and it is difficult to pinpoint. With long strides the cassowary walks purposefully towards me. I take up a defensive position behind the buttress of a kurrajong tree. The bird is merely curious. It soon wanders off, picking up some fallen satinash and tamarind fruits before disappearing as silently and mysteriously as it came.

To give the cassowary time to make its way deep into the forest, I take a closer look at this special stand of trees. The bases of their trunks always intrigue me. Of the sixteen around me, a few rise as pure straight columns out of the ground. But most do not. The grey satinash, for example, massive as it is, is slightly raised on its roots as if standing on tiptoe. The bottom of the trunk itself is not actually in contact with the ground; the whole structure is supported by flying buttresses alone. The blush silky oak seems to have feet with extra long toes that grip the soil so hard that their knuckles are arched. But most wonderful are the tulip kurrajongs. Their buttresses are narrow flanges of wood that curve elegantly and sinuously as they radiate out from the trunk in all directions and to give it light weight yet strong support.

While I walk around the kurrajong a grey-headed robin sits motionless in a small shrub. Its full concentration is focussed on some drama being played out in a tangle of vines and dead sticks between two buttress roots. I hear scuffling noises and see a movement among the fallen leaves. After some minutes a very large centipede struggles out of the debris and walks awkwardly along a tiny twig about 20 centimetres above the ground. A small brown furry ball jumps up and tries to grab it. But the yellow-footed antechinus, no matter how hard he tries to jump higher and higher, cannot reach his prey. Changing tactics he races around, climbs the twigs and leaps on the centipede, biting and tearing at it. The centipede's hard shell seems to protect it from the marsupial's sharp teeth. Slowly, dragging the antechinus along, it crawls further out on the twigs. The small mammal cannot cling to this meagre support and falls to the ground. Gripping with its multitude of feet, the centipede, which is as long as its tormentor, remains motionless on the twig. But the frenetic and tough little marsupial is equally tenacious. Tearing and biting, jumping and dragging, he eventually overcomes his prey through sheer persistence. He takes his now limp victim into the protection of the vines. He crunches it up as he eats. For half an hour the robin and I watch in fascination. Is the bird waiting for an opportunity to steal the centipede?

The track takes a right-angled turn to the south. Soon I hear the rush of the waterfall. The final steep descent is not well defined and I push through ferns reaching well over my head. I emerge over a sheer cliff looking directly into an elongated rockpool. A platypus dives and surfaces, dives and surfaces as it hunts for tadpoles and shrimps. Scrambling and slithering, I reach the creek about a kilometre downstream from the confluence. Its water foaming, the creek is forced through a crevice narrow enough for me to jump over. I am now in the wide rocky bed where it is more open. The sun is hot. Upstream is a large calm pool, downstream a series of rapids leads to the falls where the white water plunges 25 metres over a precipice.

A yellow-footed antechinus leaps at a centipede.

I like it best here when the sun is diffused by high thin clouds rather than obliterated by a low dark blanket. In this season of changeable weather there is a good chance that some time during the day you get what suits you best. This is what happens now. The sun is only slightly screened. No longer is the forest a confusing mosaic of intensely lit sun dapples on a dark canvas. In the even illumination it becomes a continuum of various shades of green from the palest to the darkest.

I sit down on a rock under a tree right on the edge of the falls. I look into the plunge pool where the water boils and rushes then settles down and runs quietly and clear into a gorge hidden from view by a wall of trees. This pool is at the bottom of what looks like a crater. Sheer rock walls rise in a full circle around it to the top of the falls and in some places 40 to 50 metres above where I am sitting. The stream enters and leaves this crater, this hole in the forest, through narrow ravines. Anywhere else but in rainforest the cliffs would be too austere, too poor in moisture and soil nutrients to support lush plant life. But here I cannot even see the rocks. Enormous trees grow out of cracks in the stone. Vines cascade from the canopy. Tree ferns ring the plunge pool. One of the largest trees, another I cannot identify, leans halfway across the falls. The tips of its branches bear delicate new growth of an intense red colour. Across the crater from where I sit another waterfall, one much higher still, pours over a precipice. I can just see its whiteness glistening through the foliage. The spray of the falls makes the air constantly damp. Many epiphytes, especially delicate filmy ferns and orchids, thrive in this atmosphere.

After a while the loud, constant rushing sound of the falls becomes irksome. I walk back towards the waterhole, also surrounded by large trees. Some vines trail their tips into the pool. My favourite seat is a huge log brought down in some flood and then left stranded. Dragonflies, some red, some blue, zip around me. A saw-shelled turtle, which dived into the dark water as I approached, resurfaces and climbs back onto its rock. The clouds are getting darker.

I am soon lost in a kind of reverie, absorbed by the never-ending fascination of the tropical rainforest. At first it seems as if the plants here are rampantly out of control, that they have run amok in a wild frenzy of growth. Certainly there is competition for light and for soil nutrients, but in tropical rainforest there is space for the widest diversity of life, which grows in an orderly progression. Plants are so dominant that the first impression is that there is very little animal life. However, there are far more kinds of animals here than plants. Among the insects alone there are tens of thousands of species. Compared to the trees everything else seems small and insignificant, yet the animals play an important role in shaping the forest.

This variety and diversity of life that is rainforest once existed over large tracts of the continent, ever since flowering plants first evolved about 120 million years ago. Before that, plant life was restricted to non-flowering kinds such as mosses, ferns, cycads and conifers. But above all tropical rainforest is an expression of *flowering* plants. The rainforests of north Queensland are now all that remain of the almost limitless rainforests that covered most of eastern and northern Australia many millions of years ago. In all other areas the climate became either too dry or too cold or both to sustain them. Only this area between Townsville and Cooktown has been tropical *and* wet since Cretaceous times, that is 140 million years ago.

Among the shrubs and vines on Bulurru are some of the most ancient and primitive of flowering plants. One, a vine called *Austrobaileya scandens*, is thought to be little different from the very earliest flowering plants. This species and other ancient forms are found nowhere else in the world in such variety.

The most ancient and primitive members of typically Australian plant families such as the Myrtaceae and Proteaceae are also restricted to these rainforests. They are the progenitors of the eucalypts, bottlebrushes, banksias, grevilleas and other species which now dominate the flora of Australia's drier regions. North Queensland's rainforests can truly be regarded as the birthplace of Australia's vegetation.

While these ancient flowering plants have kept their foothold, the forests as a whole are not at all primitive. The most recently evolved plants, the most complex and 'modern', such as the orchids, are also well represented.

This diversity of plant life is driven home to me today. As I walked the kilometre or so to the falls, I tried to identify as many of the plants, but especially the trees, as I could. I am still learning and do not yet know very many names. The variety of species is bewildering and confusing. Only a few can be identified by their trunks. For most, at least a twig with leaves, but preferably a branch with flowers or fruit as well, is needed; they are up so high and so interwoven that it is often very hard to say to which tree a branch actually belongs. What makes identification even more difficult is that there is no single definitive reference book. There is no one, in fact, who can go into these forests and confidently identify each tree and vine on the spot, and there is only a handful of people who can identify the majority of them. So far about 1200 species of trees have been recognised, not all of which have been described and named. There are at least 400 different kinds of vines, and if you add the shrubs, ferns and orchids the total comes to well over 2000 species.

On this short walk and with limited knowledge I could identify 153 plants, of which 102 were trees and 25 vines. For every species I could identify there were two or three that I could not. And because of the dense vegetation, visibility is restricted to a narrow strip. Species growing just off the track I did not even see.

I use the scientific names of the plants with a certain reluctance. These names are more accurate and less confusing, but unfortunately they are sometimes less colourful, and because they are in a foreign language they put a certain barrier between me and the living, growing plants. Sitting by the pool my mind is a jumble of names: *Symplocos, Cardwellia, Cryptocarya, Aleurites, Syzygium, Macaranga, Oraniopsis, Sterculia, Tetrasynandra, Triunia, Opisthiolepis, Peripentadenia, Elaeocarpus* ... Some roll easily off the tongue and when translated make good sense or are quite colourful, sometimes even bizarre. The genus of trees named *Sterculia*, the kurrajongs, was called after the Roman god of dung and privies called Sterculius. Some kinds of kurrajong trees have a bad smell. The Dutchman's pipe's scientific name is *Aristolochia*, which roughly means 'best in childbirth' in Greek. It was so named because the shape of the flower resembles the human foetus in the best position for birth. In some cultures the plant was also supposed to ease childbirth. *Polyalthia* is a large tree with orange-yellow fruits; its name means 'many cures' in Greek and refers to the tree's supposed healing qualities. But most often the scientific name is a reference to some obscure botanical feature or the plant's discoverer.

These names do, however, help to make sense of the forest. Once you know the names of the plants, the forest is no longer a green amorphous mass. You can mentally sort all its components and begin to understand the ecological relationships. But I would much rather use English vernacular names. It brings me closer to the forest and enhances my understanding of it. To call a tree a candlenut, knowing that its seeds—the nuts—burn so brightly that they were used as candles, is of so much more interest than calling it *Aleurites moluccana*. Vernacular names, too, however, have their problems. Some, like the botanical names, have a nice ring to them: quandong, bolwarra, carabeen, tuckeroo, blue umbrella, kurrajong, watergum. Others are descriptive. The nutmeg tree bears true nutmegs and mace. Supplejack is a vine with long pliable stems. There are also stinging trees, bitterbeans, poison plums and peppervines. But most trees are named for their use, their timber—that is, their dead wood. These names are unimaginative and have no Australian flavour to them at all, referring to the resemblance of their timber to that of species from other countries. The timber merchant's vision of the forests is of a storehouse full of walnuts, mahoganies, beeches, oaks, alders, ashes, cedars, basswoods, ivory woods, hazelwoods, pencilwoods, silkwoods, maples and myriad others. Unfortunately these bear no relationship whatever to their namesakes in Europe and North America.

The sad thing is that seeing the trees and forests mostly in terms of their economic value is an all too pervasive attitude. Even conservationists

The buttress roots of a blush alder.

talk of the forest as a resource, of its use to humans, as if this is the main reason for its preservation. This attitude shows a poverty of the imagination and an inability or unwillingness to look beyond narrow monetary thinking. The often silly common names given to trees, seeing them as commodities to be bought and sold, and the use of clichés (cathedral-like, forest giants) diminish the forest's grandeur. They cheapen them and mask their real value as the richest expression of life on earth. Let us seize the wonder and mystery of the forests and not degrade their mightiest components, the trees, with foolish and inappropriate names.

Sitting by the pool I get carried away by thoughts like these. Sometimes I get quite angry and frustrated. Why oh why is there no widespread movement to understand these forests, both in detail and as an intricate self-perpetuating ecosystem? Does the forest speak only to our greed and not to our spirit? In bleak moments I sometimes think so. But, of course, in my reading and my conversations with people who study and appreciate the rainforests I do find kindred spirits.

I am always disappointed that the arts have so little to say about rainforest, that they have given us so little insight into what rainforests are really like, what their complexities are and what feelings and thoughts they engender. Even richly illustrated books and television documentaries only take us part of the way into the forest and its workings. Strangely, it is the scientists who express what I can only call the spiritual and philosophical dimensions of tropical rainforest. This is surprising because scientists, with their preoccupation with objective, that is dehumanised, study of parts of the rainforest rather than the whole, are not the people one would expect to articulate that the forest is so much more than the sum of its parts. But many do.

Len Webb, the father of rainforest ecology in Australia, calls tropical rainforests 'the quintessence of life's mystery and power'. He also wrote:

> *They appear turbulent and sombre, yeasty and enduring, immense, indifferent, chaotic and mysterious. Yet not so mysterious that they inhibit rational thoughts about them. They are wild for exploring and thinking, because they have a quality of richness and timelessness that puts us in touch with genesis and for which we have no vocabulary ... If we do not hear or see their silent machinery of species-making, we sense its potency.*

Marius Jacob, a Dutch forester, says this about tropical rainforest:

> *But we can only understand the forests if we accept them for what they are. No one can improve an undisturbed rainforest. It is a culmination point ... as precious, worthy and sacred as any in the universe as we know it.*

The ecologist Janzen says:

> *... the only message for a tropical biologist: set aside your random research and devote your life to activities that will bring the world to understand that tropical nature is an integral part of human life.*

My favourite quote, however, is from the British botanist E. J. H. Corner writing about south Asian rainforest:

> *I measured my significance against the quiet majesty of the trees. All botanists should be humble. From tramping weeds and cutting lawns they should go where they are lost in the immense structure of the forest. It is built in surpassing beauty without any of the necessities of human endeavour; no muscle or machine, no sense organ or instrument, no thought or blue-print has hoisted it up. It has grown by plant-nature to a stature and complexity exceeding any presentiment that can be gathered from books.*

It is in this spirit that I most often enter the forest, that I want to try to understand its workings and above all to enjoy it. I rejoice in the details, raindrops hanging from moss, a shaft of light catching a bright red growing tip, the flash of colour from a flying beetle, the iridescence of a lizard's skin, red and blue fallen fruit bobbing in the creek, the perfume of a flowering shrub, a burst of chowchilla song. And I am entranced by how all the details fit together to form the immensity of the whole.

I am so lost in thought that I have not noticed the clouds getting lower and darker. I am jolted awake from my reverie by large cold drops falling on my skin. Soon heavy rain hisses into the pool and hammers on the leaves. Within seconds I am drenched.

Day of the Wompoos

BULURRU, 28 DECEMBER

Not far from my kitchen window grows a blue quandong tree which stands head and shoulders above the forest. Its crown is only sparsely covered with leaves. As I prepare breakfast, I idly look at the tree, as I do most mornings, to see what birds might be perching there. Often I am rewarded with good views of a grey goshawk, a satin bowerbird, or one of the colourful fruit pigeons. This morning I look out in open-mouthed amazement. In the quandong sits not just one wompoo pigeon or even two or three but 15! Other birds are flying to and fro, most of them coming nearer and nearer the kitchen window; more wompoos arriving, at least another 15, maybe as many as 25. It is hard to count them as they fly from tree to tree. More than 30 of these shy and beautiful birds so close is overwhelming. I had never seen more than three or four birds together. Usually they are solitary.

They have come to feed on the fruiting trees around the house. All the wompoos in the neighbourhood seem to have discovered simultaneously that the white hazelwoods, the goya trees, and the rusty fig are in fruit. So close are the pigeons that their red eyes and bills are clearly visible. The air is

A pair of wompoo pigeons. The male bows and calls his strange bubbling call: 'bock, bock, cahoo'. The female, in response, sits upright with her feathers sleeked down.

noisy with the slap and rush of their wings and their quietly bubbling voices.

The same emergent quandong and fruit trees can be seen from my desk. Throughout the day, I keep an eye on the comings and goings of the birds. The wompoos soon fill their crops, especially those feeding on the marble-sized figs. Like everything they eat, these fruit are swallowed whole. Seeing the birds almost dislocating their jaws swallowing the figs reminds me of Bill Cooper telling me about his seeing a wompoo swallowing a banana fig—a fruit the size and shape of a man's thumb! The bird almost gagged but after several attempts forced the fruit down.

Having eaten their fill the wompoos disperse. Five or six return to the quandong where, replete, they fluff and preen their feathers and then just sit and digest. They ignore a flock of figbirds that fly in and settle all around them, calling and chattering. For a few moments the figbirds cautiously watch the fruit trees. One, perhaps hungrier, braver or less careful than the others, flies to a goya tree. She is followed by another, then another and another, till all 48 birds swarm over the laden branches. Unlike the pigeons, the figbirds stay together in a flock while feeding. The colour of the males' bright red eye patch clashes with the dusty pink of the goya capsules.

Now and again a pair of white-headed pigeons arrive. Small groups of barred cuckoo shrikes come and go. They eat mostly figs. Shortly after mid-day about 25 currawongs sweep down into the quandong tree. Large black and white birds with pick-axe beaks and seemingly pitiless yellow eyes, they swarm in aggressively, calling to each other in loud yodels and whistles like a pack of marauding hooligans. Such retiring and solitary species as tooth-bills, catbirds and trillers, which had been quietly moving through the trees, leave.

Other, smaller birds are also helping themselves to the fruit. So far the house has been my best observation post. The birds ignore me even when they come within a metre of the windows. But not all of them come close to the house, so I venture out to see who else might be there. Alarm calls go up among the currawongs and figbirds. They retreat to the tall quandong. After just a few minutes I am ignored again. Soon I hear it. A faint high-pitched 'zeet, zeet' that I can only just pick up. There are fig-parrots in the rusty fig. To hear them is one thing but to see them quite another. I look at the part of the tree where the

Barred cuckoo-shrikes feeding on rusty figs.

sounds come from and examine it minutely through binoculars. Fleetingly, as he jumps and bounds from leaf to leaf and branch to branch, I see the male, distinguishable by his red cheeks. My views of the female are equally brief. She is pursued by a churring, food-begging young. After playing hide and seek with me for about ten minutes, all three suddenly take off and, flying so low over me they almost brush my hair, they go to the other side of the house and land only a metre off the ground in a native lasiandra bush. To begin with I keep my distance. The young bird sidles up to the male, begging and begging. He soon relents and feeds the youngster, pumping mashed figs from beak to beak. Once the young has been fed, the adults turn their attention to the lasiandra fruits. These are red on the outside but black inside and very sticky. The fruit are quite tasty, but if you eat a lot of them the inside of your mouth turns black. The scientific name of the plant is *Melastoma affine. Melastoma* means 'black mouth'.

That does not worry the fig-parrots. As they busily feed I slowly move closer and closer. I can now clearly see the female. She has blue cheeks. When I am about four metres from them, the birds freeze. Motionless like that, leaf-sized, leaf-coloured and with their dumpy tails even leaf shaped, these smallest of all Australian parrots would be near impossible to spot. (Whether the fig-parrot is the smallest Australian parrot depends on how you measure size. Budgerigars are slightly longer, but weigh a few grams less.) The colours on their cheeks are only tiny patches, not enough to reveal their presence. They do not consider me much of a threat and soon slink along the branches back towards the fruit.

I go back inside where my first action is to check the quandong. Again I am taken by surprise. No fewer than 60 topknot pigeons sit preening and dozing on its branches. By the look of their bulging crops it seems they too have just raided the fruit trees.

The quandong tree, being taller and having an open crown, gives a commanding view of the surroundings. So sparse are the leaves that there is no place for a goshawk or a python to hide in ambush. It is, therefore, a safe place to land, and affords a great vantage point from which to inspect the fruit trees for possible danger. Only when the birds are satisfied all is clear do they fly down to feed.

The trees spread out below the quandong must look like an inviting smorgasbord. Both the fruiting white hazelwoods and the goya trees are numerous in all directions. To one side about 40 metres away stands the dark, spreading and heavily laden rusty fig. The two smaller kinds of trees form an interesting contrast. The hazelwood is a majestic tree of classic proportions. Its large leaves are dark and leathery. Stout limbs grow at just the right angle to give it a regular rounded shape. The blue-black fruits grow in bunches like miniature grapes. If we could give it human characteristics we would think of it as strong, solid and conservative.

The goya tree, unique to the Atherton Tableland, is wild and unruly by comparison; a gypsy. Its branches are slim and finely divided, giving the trees a bushy, untamed appearance. Where branches and twigs divide, the junctions look like knuckles; even quite young trees look gnarled. Its leaves are small and pale. At the moment the dark, dusty pink seed capsules give the trees a gloss of bright colour. You rarely see the seeds with their orange, fleshy coverings, called arils. They are eaten as soon as the capsules split open, or even before as some birds will force an entry.

The day has been overcast with occasional drizzle. Late in the afternoon the clouds part and the sun bursts through. A male scarlet honeyeater, a tiny splash of colour among the green, darts from tree to tree in search of nectar. The odd ivory basswood and umbrella tree is in flower, but their nectar is much diluted by the rain.

When it is almost dark spectacled flying-foxes, one after another, sail in on their broad wings and land in the fruiting trees. Word has got around that the fruit is on at Bulurru.

Nesting

CHOWCHILLA, 2 JANUARY

I drop in at the Coopers at Chowchilla. It is a sunny and cool morning after a showery night. We take our morning tea out in the garden. As we walk past the nest of the yellow-bellied sunbird, suspended from the verandah's eaves, one of the two young birds flies out; its first venture in to the wide world. The other will no doubt soon follow. Nesting has been successful. For the parents, but especially the female, it has been a Herculean task, if that is the appropriate description for a bird so small. The task began about seven weeks ago.

Wendy and Bill had noticed the pair of sunbirds in their garden for some days. They spent many enjoyable minutes watching the birds drinking nectar from flowers. Frequently the male, who is distinguishable from the female by his metallic iridescent gorget, stopped to sing his high pitched song. Sometimes he sang from a windowsill. The energetic birds had set up their territory next to the house. Knowing the sunbirds' penchant for hanging their nests under verandahs and awnings or in sheds and outhouses, Bill put some wire hooks in likely places. On 21 November, at 11.00 a.m., the female bird went to the wire under the eaves at the end of the back verandah. Hanging upside down from it for several minutes, she examined it closely. She must have liked what she saw for only a few moments after flying off, she returned with some spiderweb which she wound round and round the hook. Nest-building had begun. She made several more trips in quick succession with more cobwebs, the 'glue' that holds the whole elaborate structure together. The male flew back and forth with her and sat nearby while she worked. He sang a lot, which is important in maintaining

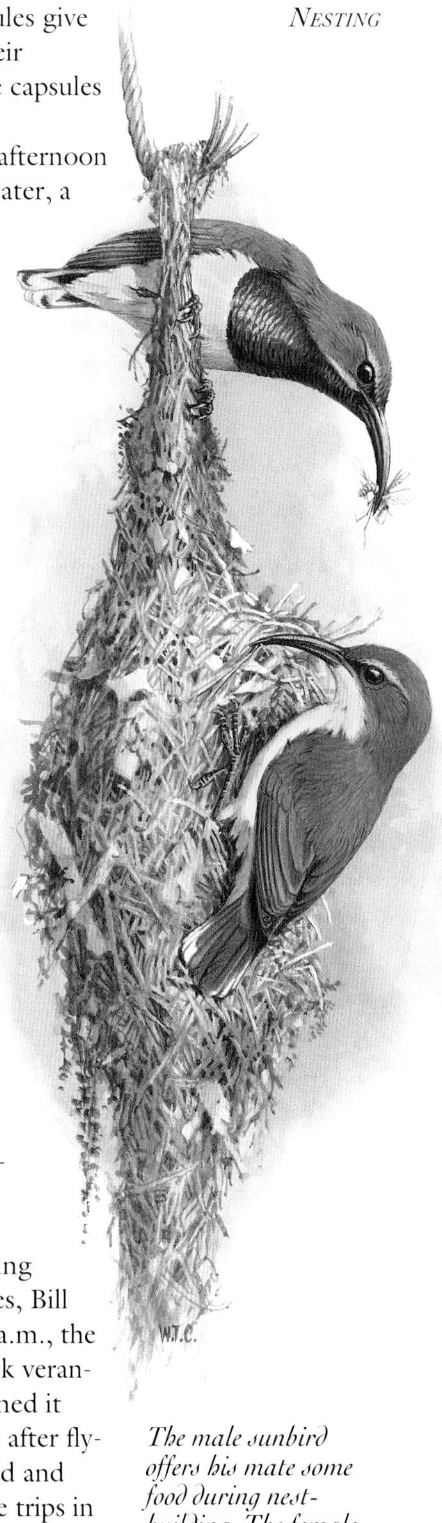

The male sunbird offers his mate some food during nest-building. The female refused.

the pair's territory, but he brought no material and took no part in building the nest.

From Bill and Wendy's point of view the sunbirds had chosen the perfect spot. They could watch the nest-building progress from their verandah.

Once the female sunbird had wound dozens and dozens of spiderwebs around the wire, she attached tiny pieces of palm leaf, bark flakes, plant fibres and caterpillar frass (a mixture of the insects' droppings and silk) to them. Flying innumerable sorties, she had woven the beginning of her nest, a piece of web and fibre 20 centimetres long and about three centimetres wide, by 24 November. Early the next day it was 28 centimetres long. When she was really busy, mostly in the mornings, she would come with material of some kind every ten to 20 seconds. Once when the female was working near the top of the nest the male landed on the wire above her. He had brought an insect which he offered to his mate. She refused it. The mass of web and fibre began to take on the shape of a nest by about midday. By late afternoon the female entered the rudimentary egg chamber at the centre of the structure. She wriggled around in it, pushing with her bill and feet. It looked very flimsy still, and it seemed her energetic kicking and shoving would cause the whole thing to collapse. But spiderweb is strong and flexible and the bird knew exactly how far she could go.

On 26 November the nest was 31.5 centimetres long and the egg chamber clearly defined. The female's squirming and fussing gave it shape and strength and made it just the right size. Over the following two days the female worked almost constantly. Twice Bill saw the male enter the nest. On 30 November the nest was finally complete. It was nearly 40 centimetres long. The female had even built an awning over the egg chamber's entrance. She would always be dry. To ensure that the eggs would also be warm she had lined the nest with the soft fluffy seeds, which look like small brown feathers, of the scentless sassafrass tree.

As if fed-up with and exhausted by her labours, the bird did not go near her nest during the next three days. Had she abandoned it after all this effort? When Bill inspected the nest on the afternoon of 4 December, however, he discovered there was a single greenish mottled egg in it. The next day there was a second egg. From then on the female, and only the female, brooded the eggs for almost 24 hours a day. She would leave the nest for only short periods to hunt for food. Every day during one of these breaks in incubation, Bill inspected the nest. On the morning of 19 December, 14 days after incubation began, both eggs hatched. The young, the size of a bean, were pink and naked. Every time their mother came to the nest, they begged for food, raising their heads, mouths agape, on trembling thin necks. Once again the female made numerous trips to and from the nest, this time with insects and spiders to feed the young. From now on the male helped a little and sometimes brought food to the nest.

The young are now out of the nest, but the female's work is not yet done. She busily gathers food all over the garden and brings the morsels to the fledgelings hiding deep in a dense clump of vines. It will be another ten days or so before the two youngsters are independent. To us this time of year, the wet season with almost daily heavy rain, seems a tough and uncomfortable period to launch offspring into the world. But birds generally are not inconvenienced by the constantly wet conditions. What the young birds need more than comfort is the season's virtually unlimited food supply of insects. These are much less prolific in cooler, drier times. Life may be damp for the young birds, but it is the best possible time to begin life on their own.

We view the successful fledging of the young sunbirds with some satisfaction. So few birds' nests that we find have this happy result. The great majority are robbed of their eggs or nestlings. Snakes, goannas, rats and especially other birds, from goshawks to catbirds, sabotage 70 to 80 per cent of all nests.

Birds invest a great deal of effort and energy in raising the next generation. Some, like the sunbirds and other small species, make elaborate nests, others, such as the rails, hardly any at all. Among pigeons and many other species males and females share the building, incubating and feeding of the young equally. The male cassowary on the other hand incubates the eggs and raises the young completely on his own while the female courts another male. Male brush-turkeys and scrubfowl spend weeks sweeping up great mounds of dead leaves and twigs which are the incubators for the females' eggs. Whatever the method or the sharing of the duties, nesting is hard work. It always seems a tragedy to see all that labour so casually destroyed in a few seconds by a predator. Yet a balance is maintained, a balance that fills the forest with birds, their colour and their song.

A pair of pied flycatchers foraging on a tree trunk.

Goshawk

Bulurru, 7 January

Loud calls of alarm come from the blue quandong this morning. A grey goshawk sits boldly on one of the exposed branches. It is quite a large individual and therefore almost certainly a female. Males are much smaller and of a lighter build.

Grey goshawk.

On the opposite side of the tree's crown a helmeted friarbird cries wolf at the top of its considerable voice. Other, smaller birds, hidden in the foliage below, twitter and scold in a warning to all and sundry. About a dozen figbirds have taken sides with the friarbird but remain silent. More and more figbirds fly in, investigating the cause of the commotion. They take the precaution of landing in a nearby celerywood. Some 70 or 80 birds, all safely out of reach, surround the raptor.

For a hunter who uses stealth followed by a burst of power and speed, the grey goshawk is unusual: its colour is in sharp contrast to its surroundings. Its white underside finely barred with pale grey, and its pale grey upper parts, afford it no camouflage. The grey colour has an unusual lavender tinge to it; in fact it was once known as the lavender goshawk. Some grey goshawks are pure white all over and are among the forest's most striking inhabitants. Whether grey or white, the birds have bright yellow legs and ceres (the patch on top of the bill), and their eyes are dark red, which all adds to their conspicuousness.

Preening her already immaculate plumage, the goshawk for a time ignores the chorus of derision. Eventually she grows tired of the incessant, noisy carping of the friarbird. Without warning she flies straight at it. The smaller bird is in no danger and takes evasive action in the forest below. Its calls, however, are more distant and less frequent.

All the other birds scatter. The fruiting trees, for once, are empty.

The goshawk, having returned to her perch, preens some more then tucks one foot under her and, spreading her wings and tail slightly, seems to doze off. But she misses nothing. Something catches her attention. She sleeks her feathers down, grips the branch with both feet and stares at something in the forest across a short stretch of grass. She bobs her head up and down and leans intently forward, to gauge the distance or get a better view perhaps. She takes off like a projectile. I see her pale ghost-like form speeding through the foliage, through what appears to me an impenetrable thicket. She knocks down a catbird, usually such a wary species, and buries her strong talons into its breast. It soon dies. The goshawk drags her kill to a large log, plucks it and begins to feed.

Rainforest Man

MURRAY UPPER, 17 JANUARY

Robert Murray is a rainforest man of the Girramay clan. He is small of stature but of sturdy build. In his mid-fifties, his curly hair is greying and receding a little. His short beard is also turning grey. For several years I have been trying to meet Aboriginal people who still know the ways of the rainforest. Their views of the forces that shape the natural environment, the plants and the animals, are very different from those of the Western culture, though equally valid. There is much to be learnt from them and I am anxious to at least have an introduction into their perceptions of the rainforest. Here, at Murray Upper, where a group of Girramay and Dyirbal peoples live, many still speak their language fluently, and still know the plants and animals of the rainforest. They could move into the forest again right now and live there in surprising comfort. Robert Murray, or Goonjurr or Dallawee, as he is known in his own language, is one of those people and I am fortunate that he

A helmeted friarbird scolds the goshawk from a safe distance.

has agreed to tell me about the forests and the way he and his people used to live there.

Murray Upper is a small community on the coastal plain. As I approach it early this morning the weather is already hot and humid. A thunderstorm has just drenched the area and is retreating to the rainforest covered hills. Around the settlement the still dripping forest is bathed in golden light. Beyond, the hills and the clouds above them are blue-black. Thunder rumbles and growls. I turn down a narrow track into an open paddock beside the Murray River. Along its bank and on the other side is rainforest. Robert is waiting for me, accompanied by his two dogs. In the shade beside the river we boil the billy over a small fire and have a cup of tea. We chat about all the rain there has been of late and speculate whether this is the beginning of the wet season. We agree that it probably is not. Robert points to a spreading tree with dark green foliage, a kind of satinash known as a river cherry, which is in flower. He calls it warrijah. He explains that if this tree is flowering profusely it means there will be a prolonged wet season with heavy rains. If there are few flowers the rains will be light. Judging by the amount of flowers out now the coming wet will be only average.

In these upper reaches the river is narrow and swift flowing. The water cascades over granite rock bars and then runs into wide deep pools. An azure kingfisher giving its high pitched call darts across the water and lands on a rock. Butterflies hover around us. The leafwing shows the orange and brown of its upper side when in flight, but at rest it becomes invisible. The undersides of its wings are perfect imitations of the dead leaves among which it lands. I notice one of its brightly coloured caterpillars on a small ground plant. Swordtails and blue triangles land in patches of damp sand to drink. Blue tigers, Australian rustics, capaneus butterflies and the occasional cruiser (butterflies have strange names) fly slowly up and down the river. My eyes are drawn, however, to the large trees growing along both riverbanks. Their branches meet in an arch above the stream. It is not only their size, though this is considerable, that attracts my attention but their rugged shapes and smooth texture. Each tree has a large rootbole folded, creased—almost tortured, it seems—into a gnarled shape. Older trees often have twisted trunks full of hollows, while the younger ones have clean straight stems. The bark is smooth, lustrous and of a pale silver-grey colour. The timber is known as kanuka box. Robert says his people call it kuddilgooroo. 'We used these for making boomerangs. We cut them out where the tree has a flange at the base of the right shape. We also use it for making nulla-nullas and spear throwers. You can even make spearpoints out of it, it's so hard.' He points to another tree nearby with a rough-barked trunk: 'That's a Johnstone River hardwood, gilla-gilloo we call it, it's even harder. You can cut good swords out of them.'

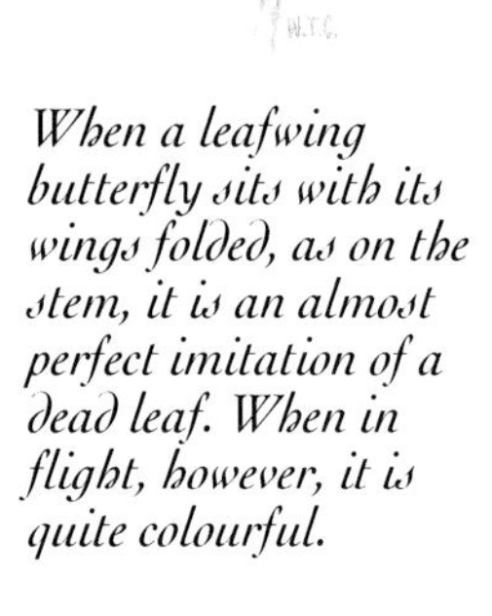

When a leafwing butterfly sits with its wings folded, as on the stem, it is an almost perfect imitation of a dead leaf. When in flight, however, it is quite colourful.

Robert douses the fire. We cross the river, the dogs running ahead. We come to a small termite mound about 70 centimetres tall. I immediately

The white-tailed kingfisher is called girra-girra by the Aboriginal people of the Girramay clan.

recognise it as the nesting place of one of my favourite birds, the white-tailed kingfisher. A pair of these birds will excavate a nesting chamber in such a mound to lay their eggs and raise their young. It is the only place these exquisite kingfishers nest. Robert knows all about these birds (he calls them girra-girra) and their nesting habits. The termites were used by the Aboriginal people too. Robert explains: 'We used these white ants [termites] to attract fish. We crushed them in a dilly bag and put them in the water. We'd give it a shake so the fish can smell them. You'd have your line in too, not with white ants on the hook, just ordinary bait. The white ants attracted the fish so they'd come closer to your hook. You can do this in still water or when it's running. Doesn't matter. We also used to drug fish,' he adds, drawing my attention to a foambark tree a few metres away. 'We'd take the bark of this tree, bash it up, put it in a dilly bag and then soak it in the water. The water got to be still though, not running. The bark cuts the oxygen off from the fish, they can't breathe. When they float on top of the water we catch them. We used to catch lots of fish here. Not much now. Not so many other animals either. I think it's because of all the aerial spraying myself.'

Caterpillar of the leafwing butterfly.

A little further on Robert stops and says: 'I suppose you want to know about this vine here.' I recognise it as the bellbird vine. 'We call it bayamudda,' he says. 'It has a long fruit that grows orange. You can eat it, and it's all right. But it comes in handy for something else too. We used to get the sap out of the fruit and used it to decorate ourselves for fighting and ceremonies. We used it like glue to stick feathers on ourselves. White cockatoo feathers are best. People used to catch young cockatoos by climbing up the tree and getting them out of the nest.' Pointing to a nearby basket fern growing on a large tree, he adds, 'We know when the young cockies are big enough to take out of the nest. It is when the leaves of this fern have completely opened out. They should be ready right now by the look of this. Cockatoos are good tucker. I should know, I've eaten plenty of them,' he says laughing.

I ask if there was a lot of fighting in the old days. 'I've only been to one fight, that's all.' He says this with some regret. 'After that the white man's law came in and they stopped Aboriginal people fighting. I was only about ten years old then, too young to fight. But my elder brothers were fighting. They used wooden swords and shields and boomerangs.'

We do not have to walk far to find trees or other plants to talk about. Just a few metres on we come to a tree I do not know which Robert calls judagula. It is yet another species with very hard wood used for making spear points and boomerangs. I ask if he used to make the shafts of the spears out of bamboo. 'No, no,' says Robert, 'we don't get much bamboo here, we used a soft wood. It was much better. We used to hunt cassowary with a spear. You have to wait for them in the right place and then corner them and kill them with spears. Same for wallabies. Sometimes we used dogs. I've hunted cassowaries a couple of times but only with a rifle.'

Still on the subject of making spear points and hardwood, Robert points to a large tree with rough scaly bark and fine small-leaved foliage which was also used for this purpose. He calls it warrigull. Its wood was favoured for making music sticks because of the clear resonance when two pieces are struck together. Robert picks up several of the tree's small seeds, which are chocolate brown, shiny and very hard. He gives them to me and says that they used these and other similar red seeds called gubbun-gubbun (coralwood) as beads to make necklaces. In the old days they pierced the seeds with a sharpened bone, but nowadays they use a heated wire. Robert says: 'The necklaces are called gooba-gooba and they can be made out of anything at all, shells or seeds. Some of them are very, very touchy. They have something to do with Aboriginal law. Only the higher blokes, the elders, like Paul Keating and John Hewson in white fella culture, know about that. It is not for young fellas. They call that gumma, it is the name for what they do. It is the same as meeting behind closed doors, no women or young fellas or children are allowed. It is only for the elders. That is what they call gumma. Whatever they talk about is between themselves, they keep it secret. It is the touchiest thing I know. I kept on asking my mother how I could find out what was going on. She told me, "You have to find out for yourself by and by as you grow up. They are not going to tell you."'

I ask Robert if he is an elder. 'I know some of it,' he says, 'but I came a bit too late for that. I am not cut or anything like that. My brother was. The last ones went through [initiation ceremonies] in about '40 or '41. I'm unlucky I missed out on all that.'

Robert walks a few paces into a patch of undergrowth and stops to pat the bark of a tall tree. 'This is gullumboodor,' he says of yet another tree I do not recognise. 'We cut bark off this to make a roof for our gunyahs, or midyahs as we call them. We used to cut pieces of bark off the big trees, but never right around them so they would die. We only took strips which grew over again. When it is fresh you have to flatten it out and put something heavy on it till it dries, otherwise it curls up. That tree over there is a yellow jack, or giddul as we say. You can eat the fruit of that, but mostly we used to cut the bark to make canoes. You need big trees for that. They've all been cut for logging. And there is another tree we used the bark of,' he says, pointing to a

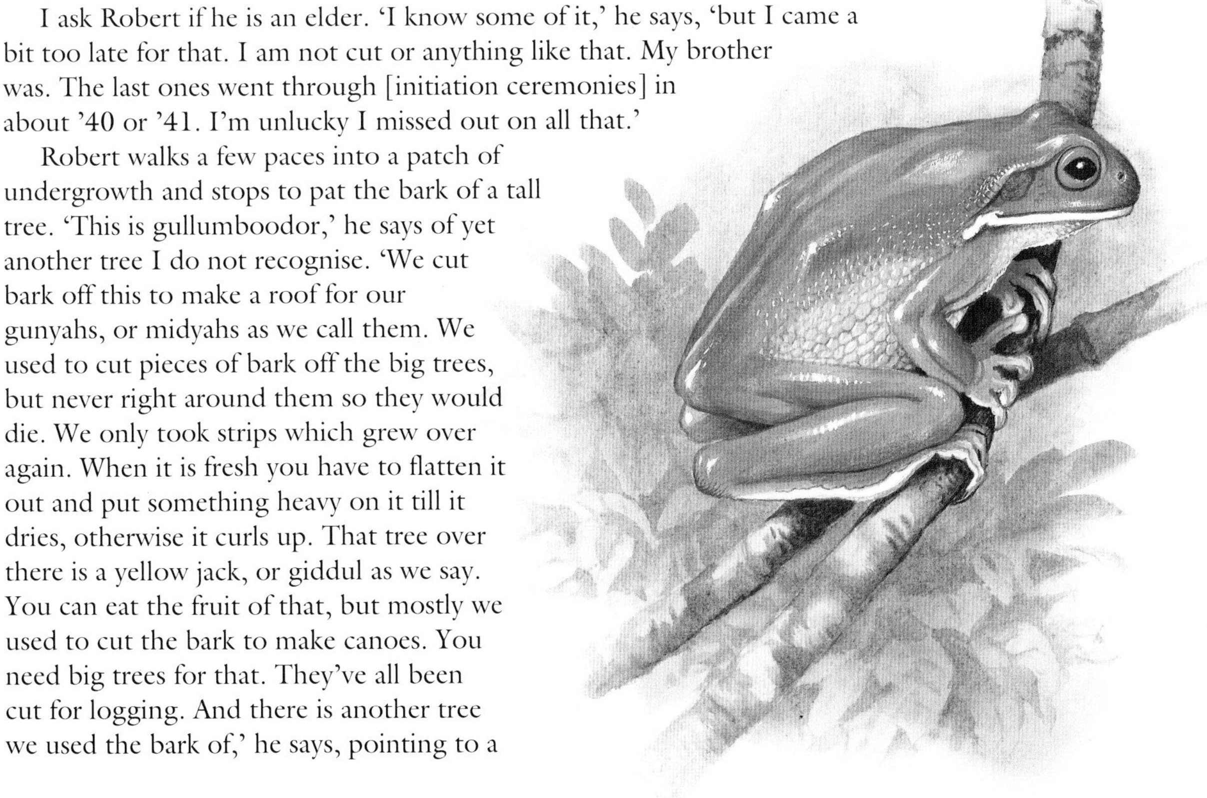

The Girramay people know the giant tree frog as bungkwee.

blush touriga, a medium sized tree with yellowish bark. 'That is nowbah. We used it to make water carriers. We get a piece of bark, scrape it, then fold it down the middle. You need to heat it over the fire to bend it or it'll crack. A sharpened wallaby bone was used like a punch and needle to make holes down the sides and then we would sew it up with split lawyer cane. The holes were sealed with the gum of the gummor tree and with beeswax. It was quite waterproof. You could carry water or honey in it.'

The sky clouds over and the rumble of a thunderstorm comes nearer. A giant tree frog, a lovely green species with a white lower lip and yellow eyes, calls loudly from a tree hollow, 'quark, quark, quark'. I ask Robert if he used to eat frogs. 'Yes,' he says, 'we eat frogs. You have to find a special one though, a brown one from near the river. We call that frog widgenburra. But the one you hear singing now, that is bungkwee—the green frog. We don't eat them. We eat tadpoles too. When there are lots in the water we catch them, tie them up in a ginger leaf to give them some flavour, and cook them in the ashes.'

Earlier, Robert had shown me how the long broad leaves of the common native ginger were used to wrap all kinds of food during cooking. He said the plant's bright blue fruit are also edible.

Some of the Girramay names for the trees are difficult for me to get my tongue around and even harder to spell. The best I can do for the northern tamarind is njohgah, but it is an approximation only. My attempts at pronouncing the word cause Robert much amusement. He points out, however, that the tree has finished fruiting and that the fruit is sour. This prompts him to remark that the Aboriginal calendar may not have been divided into months and weeks but that they knew very well what was going on in the forest at any given time. There were signs all around that told them of approaching seasons and where they should move and when to harvest particular plants and animals. 'For example, when these njohgah fruit are ripe then the scrubfowl are ready to lay, and when that vine there, we call djungeen [October glory] has ripe fruit on it then the turkey nests have eggs in them. We know when the wet season, which we call murradjah, is coming. It's when the river cherry fruit is ripe. When a certain kind of locust is singing out we know the nuts on the brown pines up in the hills are ready for us to eat. All the time we look for signs, we call them oyooroh. People think we just go walkabout, that we go out and don't know what we're doing. But we follow the seasons. We know what we're doing all right.

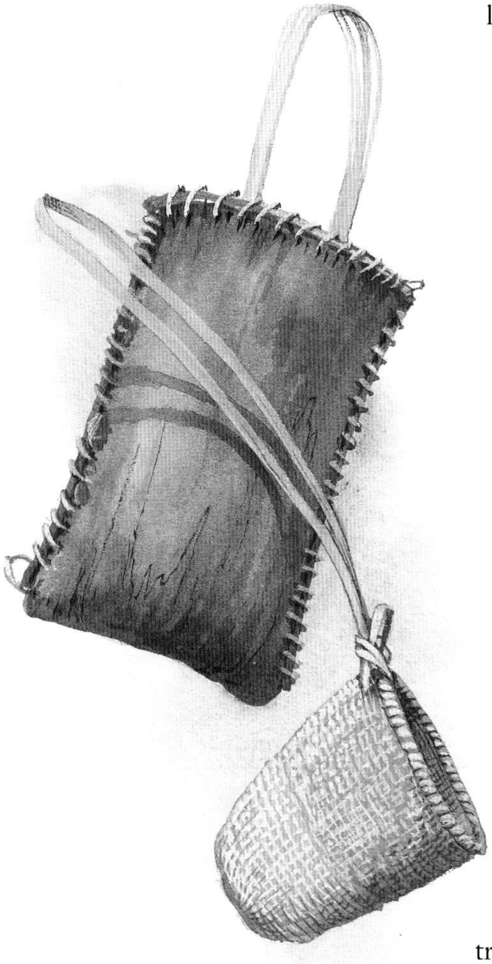

The larger container is made of the bark of a nowbah or blush touriga tree. It is waterproof and was used as a bucket. The small dillybag is made of grass stems.

'We had medicines too,' he adds. 'This one here we call it goodoombah [native nutmeg]. We get the sap out of the bark. It is reddish in colour and cures sores and ringworms. But this tree,' he says of a tarwood growing beside it, 'is a bit risky and dangerous to handle. If you get the sap on you it burns your skin. You get blisters. I got burned once when I was cutting a log. I got a big sore on me; even got compo for that. I don't like the smell of it either, and when you use it in a fire the smoke burns your eyes.' It appears a harmless enough tree with luxuriant, large-leaved foliage.

More benevolent and useful is the fig tree. 'This fig tree is what we make rope and string out of. We cut the young shoots, strip the bark off them, let it dry and then get the inner bark out. Another fig we use for making string grows at the side of the river, we call that allbooloo. This one is boonbull. We make the string by rolling the fibres on our legs. It is strong and we use it for anything at all; fish net, fishing line, turkey trap. We call that moongool, the string. When we make a fish trap we call that moogarong and when it makes a turkey trap it is called wahludgee. Three, four names all in one, eh,' Robert adds, laughing at my bewilderment. I realise that everything in the forest is finely observed and that everything has a name. Explanations and descriptions in Aboriginal languages are always precise.

While we stand quietly beside the fig tree a white-tailed kingfisher lands close to us. It sits there briefly, a green insect in its beak, then is off again; a brilliant apparition flying through the dark green forest trailing its long, long white tail. The bird does not fly far. 'He probably has a nest with young there in the white ants' mound,' Robert observes. We walk quietly towards it. Robert is right. We watch the kingfisher enter a hollow in the small cone-shaped mound, still carrying the insect. It soon reappears without it.

Everywhere in our wanderings Robert points out trees that have edible fruit: ooray, the Davidson's plum; doogooroo, the white apple or bumpy satinash; dubboguy, the powderpuff lily-pilly; gibbor-gibbor, the Herbert River cherry; mahrray-mahrray, the acid drop fruit or zigzag vine; and many more. No matter what the season there is always some fruit to eat. But you have to know what you are doing. You cannot pick up just anything that looks appetising and eat it. The fruit of the pothos vine, which he calls gooyour, Robert warns is poisonous unless treated. 'When they are ripe, they are bright red and grow in bunches like grapes. You pick them and cook them in the ashes. They're good then but if you don't cook them you get crook.'

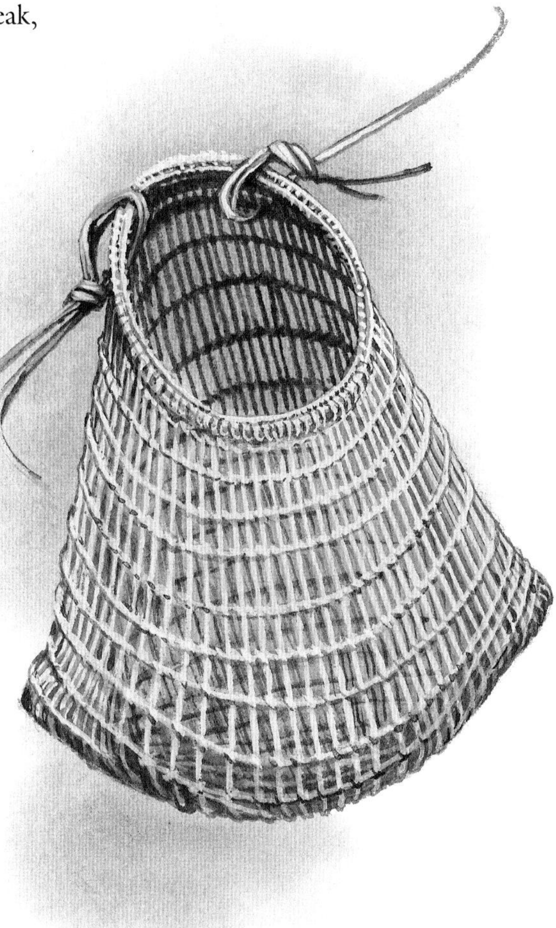

A dillybag made out of split wait-a-while canes.

One of the most common trees in the forest is the black bean, which Robert calls mirrundj. The seeds, which grow in large pods, are about the size of a small potato. They look good to eat. 'You can't eat them like that,' he warns. 'You got to cook them first and then grate them, put them in a dilly bag and soak them in running water for two or three days. You wouldn't want to be hungry, eh. But if you don't do that you've got trouble. We cut the beans with a shell.' He looks around and picks up the empty shell of a landsnail. 'We call this shell gudjeree. We break off the lip like this, so that it's sharp, and slice the beans. These shells have a stripe on them, see. Young men were not allowed to eat them. Only old people. That is Aboriginal law. It's a bit hard to explain. We still have to show respect for Aboriginal law. The young people don't know, they are not interested. We try to teach some of them, but they have a long way to go. It's not easy. I know these things because I've been taught ever since I was small.'

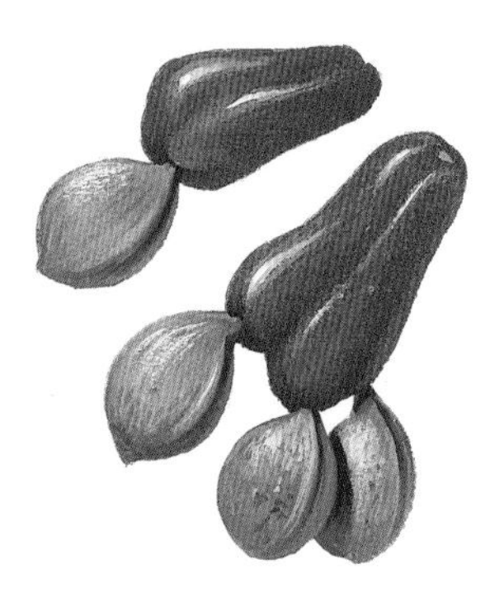

The fruits of the brown pine were highly prized by the Aboriginal people. The red segment is not part of the fruit. It is a modified stem.

Our wanderings take us back towards the river and Robert springs a surprise on me. I thought we had been walking aimlessly along, but all the time we had been making our way to this special place along the river. Up on the bank is a recreation of a traditional Girramay camp. Centre stage stands a well constructed hut of bent saplings tied together with split wait-a-while (lawyer cane). The frame is thatched with wait-a-while leaves. Beside it a stone oven has been dug. On the other side stands a frame on which fish, cassowaries or wallabies used to be smoked. A little further on a larger shelter, or midyah as Robert calls it, has been constructed. It is covered in bark, has an awning and looks totally waterproof. In front of it lies a grinding stone and several stone axes. From its front hang various kinds of dilly bags and a bark water container. The dilly bags, especially, are beautifully shaped and intricately woven, some from split wait-a-while and others from grass. The larger dilly bags, Robert tells me, were used by the women to carry their babies. All around the midyahs the ground has been swept clean to expose the sandy ground. It looks comfortable and eminently habitable and from the little I have learned food was plentiful and appetising. There may have been times when it was wet for long periods or when mosquitoes and marchflies were troublesome, as they are today, but the people wanted for little.

Seeing this camp, even if it is only a reconstruction to show visitors and the young people, and talking to Robert, it has become inescapably clear that this rainforest is the Girramays' home. This is the place they know in minute detail, a place where they are at ease, they feel connected to. The Girramay, and other traditional rainforest Aboriginal people, know the forests intimately as white people never could. Rainforest supplies not only all their material requirements but also their spiritual needs. Robert is reluctant to talk about that. Much of it is secret and he does not want to betray the confidence of the elders.

Robert shows me all the details of the midyahs. He names every variation in construction and every piece of building material. He explains how one kind of lawyer cane is good for tying the frames and another is not. 'It is the same as building a house,' he says, 'you get to know what materials to use and they all got a name. Same here.'

I ask what he used to do when it rained and rained during the wet season. 'You have to do the best you can,' he answers. 'We always had a fire because we know where to find dry wood. To make a fire in the rain you have to get some dead lawyer cane. It is very hard so you have to smash it up with a rock or something. Inside is what looks like sawdust and it is dry. We can make a fire with that, but without it there is no chance.'

As we go slowly on I ask Robert to tell me a little about his early life. 'I have lived here all my life except for a few years when I went to school in Townsville. I was told I could keep my language if I wanted to. In other places our people weren't allowed to speak our language. But even if I hadn't been allowed to speak it, I would never forget it. People who come here from other countries don't forget their language.

'When I was small I lived in the rainforest with my family. We lived on bush tucker most of the time then. We were taught how to hunt and how to survive. I really liked going out hunting with the other boys. That was a good time. Not like these days. Most young people have fallen out of that. It is a bit of a waste. In the old days most of the land was scrub [rainforest]. Then in the early '60s all of it was taken up, mostly by King Ranch. They came in with a law that we couldn't go on our own land. It was really bad to cut down all the scrub and forest [eucalypt and paperbark woodland] because now we have no place to hunt. Even the waterholes were destroyed and nearly all the scrub and forest is gone.'

We walk along in silence for a while. Robert is a bit thoughtful but soon recovers his good spirits. He takes me over to a fan palm. Folding the huge leaf in a certain way and tying the ends he makes a container. 'You can turn the leaf up like that, tie him up. Then you crack an egg of a scrubfowl or turkey into it and cook it, leaf and all, in the ashes. Or you can wrap fish up in it to cook. We also use the leaves to make a roof for a midyah. We call this palm jugoodor.' At the base of the palm, where the fronds unfold, is some natural soft material that looks like woven fabric. Robert calls it eemahn and says, 'You can wrap eggs in that and put them in the dilly bag. Saves them from cracking.' He continues, 'This is djinmaheen, the tree we make fire sticks out of. Here on this doogooroo tree you may see little black ants we call mummoo. What we do is, we cut ...'

Flamboyant and Outrageous

BULURRU, 21 JANUARY

I have just come back from a long walk in the forest where I picked up every kind of fruit or bunch of fruits I could find. I have put them all out on the kitchen table and they cover the whole of it. There are some 40 different kinds of fruits, and between them they make up as vivid and resplendent a collection of objects as you could assemble anywhere. Colour is what first catches the eye, as it is meant to do, for that is what attracts the birds who will eat the fruit and so disperse the seeds carried within them. But shape and texture are also intriguing. Glancing at the fruits, glowing in the strong light streaming in through the windows, the reds, oranges and yellows predominate, with a fair sprinkling of purples and blues.

Pride of place goes not to a single fruit but a large bunch of a hundred or so shining bright blue ones with a touch of iridescence. Each fruit is about 2 centimetres long. They come from a small slender tree in the forest understorey called blue nun or delarbrea. The dark red berries of a vine called *Hypserpa laurina* also come in bunches but with only about 15 in each. A twig of the white bollywood bears several scores of fruits of different colours, the size of currants, packed closely together. There are yellow, red, deep maroon and black ones all mixed up together, the colour depending on the state of ripeness. The black ones are the ripest. A leafy twig of a bridal bush has a small scarlet fruit in each leaf axil. The northern white beech has bunches of shiny pale purple, or is it blue?, fruit the shape of miniature apples. Crimson berry, a large shrub, contributed bunches of brilliant pink-red fruit, while the Topaz tamarind has strings of rose-red fruits which have split open to reveal polished black seeds. The handful of fruits, each a little smaller than a cherry, that I picked up from under a paperbark satinash, make a splash of vivid purple. Delicate among the blazing colours are the pale pink fruits of the small tree *Fontainea picrosperma*. They look like small peaches complete with a covering of fine white hairs. Two figs are in fruit. Those of the rusty fig are the size of marbles and orange-brown, while the banana figs indeed look like small bananas in shape and colour. Shining almost incandescent orange-red fruits come from the spice bush, sometimes called the red rattle as the seed rattles loose inside the round fruit. Red fruits like beads on a string belong to the palm-lily. The common wait-a-while is represented by a bunch of fruit almost as large as that of the blue nun, with each individual fruit much paler and more subdued in colour. They are pale yellow and their outer skin is covered with overlapping scales, like that of a snake. Each scale is edged in brown. Pink tamarind fruit, apricot coloured and apricot sized, have split open exposing black seeds partly covered in a fleshy substance called an aril. The Boonjie tamarind goes one better in this business of a covering aril. Its large fruits, about the size of a small apple, are mustard yellow in colour. The ones I found today had all split open and spilled their two or three brown shiny seeds. Each seed is wrapped in a succulent orange-red aril. They looked magnificent among the ground cover of dead leaves. The arils tasted refreshing, something that the musky rat-kangaroos and pademelons appreciated. They had eaten most of them.

Single large fruits also occupy my table: bright yellow ones of the ivory wood; blue Davidson's plums with edible, if somewhat sour purple flesh, as large as any plum you buy at the greengrocer; poison walnuts add a splash of red. But again and again my eyes are drawn to the fruits of the buff quandong, a large tree, and ochrosia, a small shrub. They are the most vivid, purest reds of all. They are not very large fruits, but in the case of the buff quandong carpet whole areas of the forest floor with their brilliant colour. They are great favourites of the cassowary.

The fruits' colours are often so subtle and so varied that it is almost impossible to describe them accurately. How many ways are there of saying purple, or red, or orange, or blue? Every imaginable variety of these colours is present in rainforest fruits. This often poses difficulties for Bill in his paintings. He says there are some colours, such as the hot pink vitex fruit, that he just cannot reproduce as brightly or as accurately as he would like. All the colours of the plant kingdom seem to reside in the fruits of rainforest plants, from the soft and subtle to the flamboyant and outrageous.

The fruit of the blue nun or delarbrea grow in large bunches.

Not all the fruits on the table are brilliantly coloured. Among them is the long oval shaped woody capsule, about 12 centimetres long, of the northern silky oak. It is brown-grey in colour with a soft velvety texture. Beside it lies the smaller green capsule of the brown silky oak which contains seeds with transparent wings. When the capsule splits open these miniature propellers fall out and, whirling about, slowly drift down, winging their way some distance from the parent tree.

When you consider that this collection was gathered on just one brief trip into one type of forest at one particular time of year, you can get some idea of the great variety of fruits in the forests as a whole. During every month of the year there are fruits ripening somewhere. Every month there is some fruit that through a combination of size, colour and abundance makes you rub your eyes in disbelief. The 25-centimetre-long red and yellow spiralling pods of the scarlet bean are on in November. In October the brilliant red and glossy fruits of the scarlet satinash, as much as ten by six centimetres in size, fall in great profusion. Cassowary plum trees are hung with large red-purple fruit in March. In July the mountain gardenia has large fruit of an indescribable orange-pink, while the yellow ovals of the walnut *Endiandra montana* rain down in June. And so it goes on and on. Always there are new surprises of brilliant colours shining in the green foliage or scattered on the forest's brown carpet. An abundance of colourful fleshy fruits is one of the characteristics that distinguishes tropical rainforest

Fruits of the forest, from left to right: the walnut Endiandra montana, *buff alder, scarlet satinash,* Dichapetalum papuanum, *the pink* Fontainea picrosperma, *and the blue cassowary plum.*

The fruits of the Boonjie tamarind contain brown seeds covered in orange-red, juicy arils.

from all other habitats, even subtropical and temperate rainforests. It is the fruits rather than the flowers that bring colour to the forest.

A plant's fruit is the package that contains its essence, its seed. These packages, besides being conspicuous, often taste good and are attractive to many animals. Not all fruit, however, can be eaten by all animals. People, especially, are excluded from enjoying many of the rainforest's berries, 'plums', 'apples' and nuts. As there is as yet little knowledge available as to which fruits are edible and which are poisonous, it is a good idea not to try any, no matter how tempting they may look. A forest ecologist working specifically on fruit was in the habit of tasting most fruits he came across. He would take just the tiniest nibble, test it for flavour, then spit it out again. He tried that with the spice bush fruit. Not long after tasting it he collapsed with severe stomach cramps and intense nausea. He had to be hospitalised for a short while. Yet some of the forest rodents will eat this same fruit.

It is of no benefit to a rainforest plant, especially a tree, to have its seeds fall immediately below it and stay there. The few seeds that would germinate would not grow to mature trees in the dense undisturbed forest. The seeds need to be spread over as wide an area as possible so that at least some can grow into large trees. Regeneration in rainforest is difficult enough, as seedlings can only grow where there is a break in the canopy, in a place where a tree has come down in a high wind or has otherwise died and collapsed. Seeds have to find their way to such places. Some drift in on the wind but most are transported by animals. The chief dispersers of seeds are birds, with mammals playing a minor role. In Australia reptiles and fish are not known to disperse seeds. About 40 species of birds have rainforest fruit as a major part of their diet. They digest the outer covering then pass the seeds, which are often protected by a hard indigestible coat. The birds can cover considerable distances between eating the fruit and voiding the seed.

This system of dispersal works perfectly for small-fruited plants. Their seeds find their way to openings in the forest and cleared spaces with the help of birds and, to a lesser extent, flying-foxes. The bats tend to spit out the seeds while feeding and so are of no great help.

But what of the large-fruited species, the big 'plums', satinashes, walnuts, silky oaks and certain gardenias? These kinds of trees occur throughout tropical forest in considerable numbers and variety. For them the cassowary is the main agent of dispersal. The size of a fruit that a cassowary can swallow is phenomenal. It has no trouble with a good sized apple, and few rainforest fruit are that large. Cassowaries are rare now and have disappeared from large areas of their former range. Will the large-fruited trees now also disappear from those places? As they grow to be many hundreds of years old, it is difficult to be sure in the short term. Records for most rainforest trees go back for a lot less than one hundred years. Recent studies have revealed that some mammals may help disperse large fruits. White-tailed rats and musky rat-kangaroos will cache large seeds.

Red rattles, fruits of the spice bush, lie like jewels on the forest floor.

These caches are sometimes forgotten and the abandoned seeds will germinate if conditions are right. Perhaps the large-fruited trees are a remnant of the times when moa-like birds, very much larger than the cassowary, roamed in these parts. In terms of the great age of the rainforest that is not so long ago.

Whether large or small, birds and fruit are closely tied together. Many seeds do not germinate unless they have passed through the alimentary tract of a bird. The bird's droppings, especially in the case of the cassowary, also provide fertiliser to speed the seedlings' growth.

The fruits look good, they mostly taste good, but as a food they do not rate very highly. They are low in protein and high in carbohydrates and sometimes fat. Few birds and mammals eat nothing but fruit. Among mammals, flying-foxes come closest to an exclusively fruit diet, but even they will switch to nectar and pollen whenever the opportunity arises. Riflebirds and bowerbirds eat considerable quantities of insects as well as fruit, as do honeyeaters and starlings. The cassowary eats fungi and any dead animals it can find. Only the fruit pigeons, the wompoos, the purple-crowneds, the topknots and the nutmegs, are exclusive fruit-eaters. To get sufficient protein, they must eat enormous quantities.

Fruit is not a good food to bring young up on either. The smaller birds all raise their young on insects and spiders. Fruit pigeons, like all pigeons, produce a milk-like substance in their crops and feed their young on that. Among mammals, of course, this is not a problem as they suckle their young.

While the fleshy outer covering of the fruit is a poor food, the seed is a different matter altogether. It is often very nutritious indeed. Many native rodents have discovered that and live on a varied diet of seeds and nuts. Among birds, parrots and cockatoos are also aware of the difference and will open the fruits, pierce the seed coats and eat the kernels. Sulphur-crested cockatoos have such strong beaks that they can bite through the thick woody capsules of the northern silky oaks.

For the plants that is disastrous. Parrots and rodents have in effect ignored the packaging and penetrated the seeds' defences. Each seed-eater is a destroyer of potential plants. Luckily, seed-eating birds are few in number in the rainforest and while feeding they spill, and therefore broadcast, a great many seeds.

Butterflies Dance and Frogs Sing

Bulurru, 8 February

Quite unexpectedly the heavy rain stops. The clouds disperse. Hot afternoon sunshine sparkles on the dripping foliage. Birds burst into song. Tiny insects dance in clouds above some shrubs. Larger ones buzz lumberingly by or zip through the trees in fast flights. The shrill songs of cicadas hammer against my ear drums. But none seems to be as liberated by the sunshine as the butterflies, and February is the month when butterflies are most numerous.

Down the slope to the west of the house I have left a large patch of lantana. These shrubs, which sometimes erupt into vines, were introduced from Mexico many years ago. The lantana's vice is that it invades cleared spaces and grows into such impenetrable thickets that no other plants can get even a toe hold. But they also have a virtue: butterflies love the nectar of their prolific flowers, and what is even more wonderful is that they entice many species down from the tree tops to eye level.

The most striking of the canopy butterflies is the Ulysses. One of them hovers close to my face as he dips into flower after flower with his long, black, roll-up tongue. He does not land as the lantana flowers cannot support his weight, which is my good fortune for if he could, he would fold his wings closed and all I would be able to see would be their black and brown undersides. Hovering as he sips nectar, flying from flower to flower, he is a shimmer of intense blue.

A pair of birdwing butterflies 'dance' in flight. The male has the green and black wings.

Smaller butterflies have no such problem; they land on the flower heads, fold their wings, suck the nectar from the flowers and then flutter, or dart, to the next flowers. As if in compensation, many have brightly coloured underwings, red and yellow in the union jack, vivid green in the Macleay's swallow tail, warm yellow-brown in the Australian rustic, an intricate pattern of blue, red, yellow and brown in the capaneus butterfly. A few species seem content to just spread their wings and soak up the sun. A blue-banded eggfly, a pristine individual that must have only just emerged from her pupa, sits with blue and black velvet wings spread. A cruiser, another large species, in striking orange, alights on a mossy log and spreads his wings. He is a rare visitor from the lowlands who must have lost his way.

Some butterflies are never still. Large black and white orchard butterflies flap around me. Smaller green spotted triangles, their wings a chequerboard of green and black, move so fast you have to be alert to see them. Tiny skippers and even tinier blues add pinpricks of yellow-brown and iridescent blue as they flit from flower to flower.

Into this melée of colour Australia's largest butterfly descends majestically. A female birdwing sweeps down from the canopy to sip nectar from lantana flowers about 40 metres from where I am standing. She is not alone, three males—dazzling in green, gold and black—come sailing down after her. They fly around and around the female. I cannot see clearly what they are doing. I remain standing still, hoping the scintillating quartet will come closer.

After about ten minutes the female lands on some flowers only two metres from me. She touches the flowers lightly with her long black legs, like thin wires, while keeping her balance by slowly flapping her huge wings. She unrolls her black tongue, several centimetres long, and siphons nectar from one flower after another. One male hovers with heavy wingbeats about 20 centimetres below her, keeping an exact distance as if he is attached to her with a piece of string. They fly in perfect unison. The other two males hover around the pair. The attendant male seems to be losing patience. He flies up from below the female, brushing her antennae as he rises in front of her. The female takes no notice and continues on her way. After rising the male falls a few centimetres behind then, catching up and flying below the female, he touches her again as he flies up swiftly in front of her as though to impress her with the splendour of his wings. He continues to circle the female again and again in a vertical plane. The other two males come closer and surround the dancing pair in a flurry of green and yellow. The butterflies come so close I can see the brushes of fine hairs, like sets of long gracefully curled eyelashes, on the inside of the males' hindwings, close to their bodies. These hairs broadcast a scent containing sex hormones which attract females and induce them to mate. It is these hairs that the male tries to brush against the female as he rises up in front of her. The majority of females, however, are mated almost as soon as they come out of their pupae. Males are able to detect females before they emerge and will often wait beside a chrysalis till it breaks open and the female crawls out. The ritual dance in front of me is more than likely in vain, the female almost certainly has mated already and she mates only once. She is intent on gathering nectar, the energy she needs to develop her eggs. She ignores the males, but that does not diminish their ardour. The caravan moves on. Minutes later a fourth male comes along, following much the same route as the others.

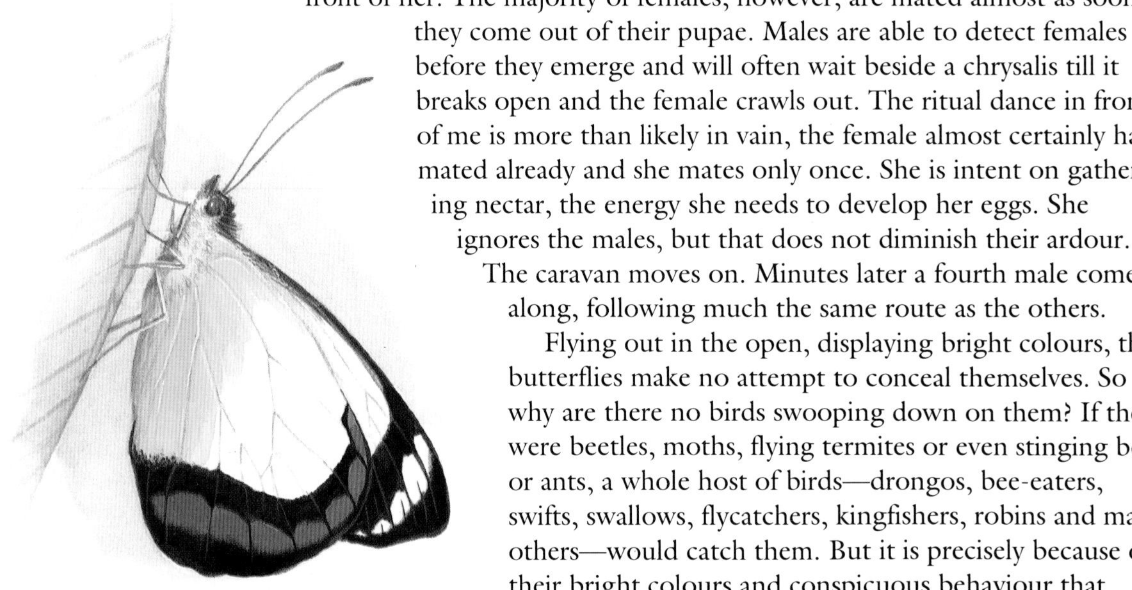

Union Jack butterfly.

Flying out in the open, displaying bright colours, the butterflies make no attempt to conceal themselves. So why are there no birds swooping down on them? If these were beetles, moths, flying termites or even stinging bees or ants, a whole host of birds—drongos, bee-eaters, swifts, swallows, flycatchers, kingfishers, robins and many others—would catch them. But it is precisely because of their bright colours and conspicuous behaviour that birds, as well as lizards and mammals, leave them alone. Bright colours act as a warning that these butterflies not only taste bad, they are poisonous. It is a lesson most predators learn quickly. Ones that do not pay the price. There is the case of an otherwise healthy, well fed bearded lizard. For four days in a row the lizard ate, of its own free will, two wanderer butterflies. On the fifth day, the lizard died. It had not heeded the butterflies' bright orange colour, warning of their poisonousness.

Butterflies absorb their poison from the plants they eat as caterpillars. Somehow these can cope with the toxins the plants produce, and store them in their tissues. One species of butterfly has caterpillars that feed on the leaves of stinging trees, perhaps the most toxic of all the plants in the rainforest. Plants are thought to produce their toxins specifically to stave off the attacks of insects and other leaf eaters. As always, there are some organisms that overcome these defences and even turn them to their own advantage. The web of interconnections and adaptations is endless.

The sunshine is short-lived. Low clouds roll in from the northwest. Rain threatens. I take shelter under some trees and soon hear the intermittent plop, plop, plop of the first heavy drops hitting the leaves. Faster and faster the drops fall till they create a roar that drowns out all other sound. The rain is heavy, the clouds are dark and low and come from a northerly direction. The monsoon is here at last.

After dark the heavy rain changes to a misty drizzle. In the gully directly below the house a spring trickles into a small circular pond overhung with cunjevois, ferns and shrubs. The day's rain has filled the pond to overflowing and drips from the leaves. Frogs love this place on warm wet nights of gentle rain. Tonight they have gathered there to sing. Carrying a torch and sheltering under a large umbrella, I too am drawn to the spot to investigate the chorus of peeps, quacks and trills.

The red-eyed tree frog sings from leaves hanging over the pond.

My arrival silences the singers, but not for long. Soon after I have switched off the torch a small frog gives a tentative peep in the leaf litter behind me. It is answered first by one then another and another in a chain reaction of soft peepings. From the leaves over the pond comes a much louder, drawn out 'wah, wah, wah', beginning slowly and softly, then gathering momentum and volume. Again the answer is swift, and now deafening, as a frog less than a metre from my ear calls at full volume. Soon a dozen sing over the pool. When they reach the final, loudest 'wah', they stop, as if gulping for air, and call more softly 'cree, cree, cree'. Once recovered from the effort the chorus starts again, the volume increasing as the singers urge one another on.

Around the edge of the pond other frogs call irregularly in deep-voiced single notes of 'wahk', while from the water itself come sounds of 'toc, toc' like miniature muffled explosions. Occasional bursts of rapid quacking are fired across the pond. It is pitch dark and I can only hear the frogs. I wait till the chorus is in one of its rising cycles of passionate song and switch on the torch. Most of the frogs are too absorbed in their singing to stop. The light catches a brilliant green frog, about six centimetres long, sitting on a leaf only about two metres away. His throat is inflated and he calls 'wa-a-a-h' till it seems he must burst. His underside is yellow and he has orange eyes. In spite of that he is known as the red-eyed tree frog. A dozen others around him also strain their voices, till one after another their balloon throats slowly deflate with calls of 'cree, cree, cree'. These songsters are all males trying to attract females to the pond with their far-carrying voices. As always I am astounded that such small animals can produce such an enormous volume of sound. It is made possible through their inflated throats which are efficient amplifiers.

I switch the light off. When the chorus once more throbs with ardour I switch on again and try to find some of the other singers. The 'wahks' come from large frogs sitting among dead leaves on the edge of the pond. They are easy to locate for their huge eyes reflect the torchlight. If they had not, these northern barred frogs would be almost impossible to pinpoint. The pattern and colour of their backs is a perfect match for the dead leaves on which they sit.

The quacking comes from a broad-palmed frog sitting on a small stone. It is one of the rocket frogs, so called because of its pointed snout and powerful hind legs which propel it like a rocket. The 'toc' sound is more difficult to trace, but eventually I track it down to a mossy overhang which hides a striped marsh frog.

The barred frog is perfectly camouflaged among the dead leaves of the forest floor.

I know the peepings are made by a tiny frog, no more than two and a half centimetres long, hidden in the leaf litter. It could be one of a number of species of frogs in the family Microhylideae. All are small species very different from Australia's other frogs and, except for one, restricted to tropical rainforest. I can hear them everywhere. Time and again, by homing in on the calls, I think I know exactly where one is hiding. But when I look I cannot find it. After many minutes, slowly moving closer and closer to a particular voice, I am pretty sure its owner must be in a patch of litter no more than half a metre square. Carefully I lift one damp, dead leaf after another, examine it and place it in a pile. When I am almost down to the last leaf, still without finding the frog, I hear 'peep' from the leaves I thought I had examined so carefully. I start again in the reverse order and finally discover the frog, a small brown smudge on a brown leaf. No one knows exactly how many species of microhylid frogs there are; it may be six or it could be double that. They are very difficult to tell apart. To be absolutely sure as to which species it belongs you have to dissect the frog and examine its skeleton. I am not about to do that, and after all my efforts I return the tiny amphibian to his damp retreat. Microhylids do not go through a tadpole stage in water. They lay their eggs in damp soil and when they hatch the tiniest imaginable, but fully formed, frogs emerge.

Not so for those calling at the pond. They must lay their eggs in or near water. The males' songs have attracted some females. Pairs of red-eyed tree frogs float on the pond's surface, the male clasping the female from behind. While the female lays the eggs he fertilises them. The clear eggs with dark centres will remain floating in the water till the tadpoles wriggle out of them. A female northern barred frog is held in a similar embrace by her mate and lays clear, shining, pea-sized eggs on a large dead leaf just above the pond's water level. When the tadpoles struggle out of the eggs they will fall into the water. If more heavy rain were to fall, which is more than likely, the eggs and tadpoles will be washed into the pond and may even be swept many kilometres downstream.

This is not the frogs' first breeding of this wet season. Only a few females have made their way down into the gully. The males without partners continue to call and call. Their voices ring in my ears as I make my way back to the house.

What is a Rainforest?

BULURRU, 10 FEBRUARY

The monsoon has settled in and it is still raining. Sometimes the rain falls as a light drizzle, but today it comes down heavily. Hour after hour it drums on the iron roof. The overflow from the water tank gushes out over the lawns. In the forest, rivulets stream down the tree trunks and disappear into the leaf litter. No bird sings nor moves about. Insects hide under leaves or creep into narrow crevices. Musky rat-kangaroos curl up in their nests; pademelons sit miserably under small bushes. Giant earthworms, of a strange blue colour and half a metre or more in length, are flushed from their underground tunnels and crawl awkwardly among the dead leaves. Platypus stay in their burrows. Animal life has shut down, is holding its collective breath, till the rain eases. The plants, on the other hand, thrive.

All day the light is an even dull gray. Ragged swirls of mist drag over the tree tops and drift into the valleys. Sometimes the rain is so heavy I cannot see the quandong and fig trees close to the house. Moisture envelops everything, seeps into every corner. It is not a time to be out in the forest.

The enforced confinement to the indoors is a perfect time to consult my library in my continuing search for the rainforest's identity, its essence. What really is a rainforest? What conditions does it need, and how does it differ from other habitats? The unique structure of the rainforest, and the inter-relationships between its many and varied plants, are often lost sight of when you are walking about, looking at particular plants or being enthralled by the animal life. How does it all fit together? Is it just a wild, uncontrolled, intertwining mass of rampant vegetation, as it sometimes appears, or is there some kind of order at work?

As their name suggests, these forests need a copious supply of rain, but that in itself is not enough: the rain must be more or less evenly distributed throughout the year. For *tropical* rainforest, and that is the only kind we are concerned with here, to flourish the climate must also be warm. The most complex, the ultimate rainforest, also requires good deep soil. In other words the 'best' rainforest, that is the most complex, with the tallest trees bearing the largest leaves, with the most vines, ferns and epiphytes, grows where there are virtually no limits to plant growth; where there are no frosts, no prolonged dry spells, no soil infertility to eliminate sensitive species. Luckily Bulurru, with its high rainfall and deep basalt soils, falls within this category. However, conditions that vary only a little from the ideal still support rainforest. Cool mountain tops, the drier western edges of the tablelands and poor soil all carry rainforest, but they are less complex, they are not quite the ultimate.

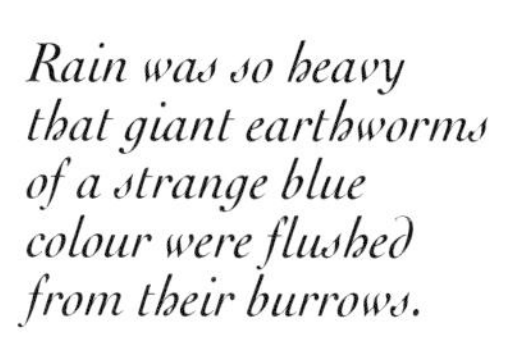

Rain was so heavy that giant earthworms of a strange blue colour were flushed from their burrows.

Tropical rainforest needs at least 1500 millimetres of rain each year (Bulurru receives 4020 millimetres annually) without any rainless months. The monthly average temperature must be about 19°C, without great daily fluctuations. Because of the rainforest's efficient nutrient turnover, known as the litter cycle (of which more later), it is self-sustaining and can tolerate poor soils but these must be well-drained. Waterlogged soils do not support rainforest. Another element that is necessary to maintain rainforest, not by its presence but by its absence, is fire. While Australia's extensive eucalypt forests are largely shaped by fire, rainforest is destroyed by it.

Mature tropical rainforest, be it in Brazil, Sumatra, Zaïre or Australia, has a distinctive appearance. It is dominated by trees with an average height of about 30 metres. An occasional giant, called an emergent, may project another 15 metres above the general canopy. Below it are two or three more strata of trees, those which can tolerate lower levels of light or younger trees waiting for a break in the canopy to develop to maturity. It often looks as if the branches of the tall trees interlock but, because of a phenomenon known as crown-shyness, they rarely do. Each tree crown has its own space, separated from its neighbours by the narrowest of gaps. Even so the canopy shuts out nearly all direct sunlight to the forest floor, so creating two separate realms; one at canopy level, the other on the ground. Even though these two worlds are on average only 30 metres apart, they are as different as day is from night.

At tree-top level there is brightness, fluctuation, and occasional violence. During periods of sunshine the temperature may reach 33°C which drops to about 22°C at night. Humidity rises to about 90 per cent after dark but drops to around 60 per cent during sunny days. The tree crowns are frequently buffeted by rain and gale-force winds.

At ground level all is quiet and calm and sometimes even gloomy. There is little or no wind. Humidity hovers around 90 per cent night and day, and the daily temperature fluctuates by only about 5°C. Rain which lashes the tree tops reaches the ground in drips and trickles. Light intensity is only about one-hundredth of that at the canopy. There is so little light that only a few plants, those adapted to photosynthesise in perpetual gloom, can establish themselves. One of these is the fernlike cycad known to science as *Bowenia spectabilis.* Other ground plants have solved this problem by becoming parasites, that is by stealing the nutrients from other plants.

Green plants derive their energy from sunlight. Their colour is given them by a substance called chlorophyll and it is chlorophyll interacting with light that creates a tiny electrical current. The electrical energy is then converted to chemical energy which results in the growth of plants. The process is called photosynthesis. Not only are plants dependent on this conversion, but all life on this planet is ultimately driven by it.

The distinctive appearance of the rainforest, the feel and the look of it, the ambience you absorb as you walk about in it, all are largely a function of the plants' adaptations to and competition for their energy source, the light. The trees, tall, slender, without low branches, derive their shape from ever pushing their leaves towards the light. Climbing plants, strangling trees and epiphytes, so typical of rainforest, are moulded by their efforts to steal a march on their competitors for a place in the sun.

The pothos vine climbs up its support, such as a tree, with fine clinging roots.

Climbing plants are much more abundant in rainforest than anywhere else. According to some forest ecologists they define rainforest; if they are present in a forest community it is rainforest, if absent it is not. About 90 per cent of climbing plants occur in the tropics, and in some rainforests one flowering plant in twelve is a climber.

The stems of vines grow rapidly. Crawling up, twining around or just leaning on the trees, they soon reach the level of the canopy. They struggle up to the roof of the forest in five different ways. Scramblers, of which the wait-a-while is a typical example, use long whiplike appendages studded with hooks to fasten onto anything they can reach. Others just lean on and then smother their supports. Smaller, neater species, some ferns, aroids and pandans among them, climb tree trunks by gripping the bark with strong clinging roots. Twiners wind themselves around and around tree trunks till they reach the canopy. Other vines attach themselves with thin wirelike tendrils that seek out any support, even a tiny twig, and then wind themselves tightly around it.

A vine's route to the sunlight is a short-cut, for it does not have to spend many years building a solid support for its crown. But it can still be a long, convoluted journey. The stem of one climbing palm in south-east Asia was measured at over 200 metres. The average length of a mature wait-a-while, also a climbing palm, is however, less than a third of that.

Another climber makes its way up to the light by twining tightly around a tree trunk.

Once a woody vine or liane has reached the canopy it quickly grows a large crown, rivalling that of the largest trees in size. In doing so it often smothers and pushes aside the branches of the tree that supports it. Over time the vine may become too great a burden for the host tree and the entire vine and its supporting branches will then crash to the ground. Because of its flexible stems, the vine is rarely killed and will quickly grow up again, leaving a tangle of rope-like loops and knots as a legacy of its temporary setback. Old mature vines, which may have stems 25 centimetres or more in diameter, have no branches or support below canopy level. It looks as though they made their way up by some magic Indian rope trick, but what has really happened is that they have outlived their original support. However, like the saplings hanging on for years in the low light levels of the forest floor, waiting for a gap in the canopy, climbers too need strong light to make their way aloft. They cannot reach the canopy on the meagre energy of the forest gloom.

Another kind of plant has solved that problem rather neatly and ruthlessly. These trees, mostly species of fig, take a short-cut by germinating not on the ground but in a suitable place high up in another tree, where they receive enough light to grow vigorously. Strangler figs do not wait for an opening in the forest to begin the struggle upward for a permanent place in the sun. They seize the initiative, start at the top and work their way down.

The fruits of strangler figs are great favourites with birds and flying-foxes, and these animals distribute the seeds throughout the forest. If a seed happens to lodge in a place where there are sufficient nutrients, in a tree fork for instance, where dead leaves and other debris have accumulated, it germinates and quickly establishes a foothold. The growing seedling sends roots 20 metres or more down along its host's trunk. Before long these aerial roots, looking like ropes and cables, reach the soil and put on rapid growth. The seedling expands into a small bushy tree standing in the canopy. Over the years the roots multiply, thicken, join together and tightly envelop their supporting trunk. The fig's crown in time overshadows that of its host. In the end this host has its trunk completely

Basket ferns use the trees' branches and trunks for support only; they are epiphytes not parasites.

enmeshed, its leafy branches are smothered and the usurper's vigorous root system steals its nutrients from the soil. Gradually, remorselessly, the host is strangled and eventually dies, yielding its place in the sun to its assassin. The original supporting tree's trunk rots away, leaving the strangler standing like a hollow tower, a tower that may grow to be among the very largest trees in the forest.

There are several species of strangler fig. One is the rusty fig. The banana fig beloved by flying-foxes is another. But the grandest, most majestic of all the stranglers, and among the most impressive of all the trees, is the curtain fig of the Atherton Tableland. As its name suggests, the aerial roots do not merely wrap around their host but cascade down in sheets of intertwining stems, some thick and woody, others as fine as hair.

The layer of small plants that is found at ground level in other habitats has had to migrate upwards to the light in rainforest; lichens, mosses, ferns and small flowering plants are forced to grow on the trunks and limbs of large trees. These plants, like the vines, are entirely dependent on the trees but only for a foothold; they are not parasites but epiphytes. Epiphytes festoon large trees such as the emergents in luxuriant hanging gardens that may contain more than fifty species of plants.

Epiphytes usually attach themselves to their support with a network of clinging roots. This gives them a place in the light, but does not solve another problem: how to obtain the moisture and nutrients that other plants derive from the soil. A group of large ferns, known as staghorns and elkhorns, overcomes that most effectively. They have two kinds of leaves, each with a different function. One kind is broad and rounded. Several together form into a rosette around a tree branch or trunk. These rosettes grow in the shape of a basket that collects falling leaves and other organic material, which decay and provide food for the growing fern and also soak up and hold moisture. Some of these ferns, and others with a similar growth habit, grow baskets so large that they may contain half a tonne of humus complete with earthworms, millipedes and other organisms usually found in the soil. The ferns' other kind of leaves are narrow and a darker green. They carry out the photosynthesis and bear the reproductive parts, the spores.

Orchids, another major group of epiphytes, have solved the problem another way. They store water in fleshy leaves and swollen stems. Their roots swell and absorb moisture from dew-laden air as well as from the rain. They somehow find enough nutrients in crevices in bark or in the dead leaves trapped among their tangled roots.

Orchids too, like this spider orchid, are epiphytes. Only in rainforest is there such a profusion of epiphytes.

Plants that need to establish themselves high in the trees also need seeds and spores that lodge in the right places. Seeds that are heavy and fall to the ground would be useless. The spores of mosses and ferns are microscopic and so light that they are carried on the wind. Similarly, orchids produce huge numbers of tiny seeds that float on the air currents. A single orchid seed pod may contain as many as three million seeds. At least some will land in places where they can germinate and grow.

Among the trunks, flowers and leaves of the trees there are several characteristics which are found only in rainforest. The most notable one is the development of huge plank buttresses in many species. These thin slabs of wood flanging out from the base may be of a neat triangular shape stretching from the trunk to a horizontal root close to the surface, or they may curve and twist into fantastic shapes. Trees with buttresses, and not all rainforest trees have them, are shallow rooted and lack a taproot. The wonderful and often elaborate structures, therefore, give them greater stability, especially on steep slopes with shallow soils. But they are of limited help. In really violent winds, buttressed trees too are toppled.

Trees in other forests bear their flowers at the ends of their branches. But in rainforest, and only in rainforest, there is a flowering phenomenon known as cauliflory. Cauliflorous trees grow their flowers straight from their trunks. One of these is the bumpy satinash. It sprouts large bunches of white or pink flowers from its bark, from ground level to a height of 20 metres or more. These are followed by similarly coloured fruits. The first Europeans to take note of this phenomenon, in 1752 on Java in Indonesia, did not believe their eyes. When they saw bunches of flowers sprouting directly from a solid tree trunk, they declared them to be parasites belonging to a different species.

Not all rainforest trees are cauliflorous.

The leaves of rainforest trees range in size from 150 centimetres to just a few centimetres. But no matter what their size, the leaves of 80 per cent of species have one thing in common: they end in a distinctive point known as a drip tip. This construction helps to drain the leaves more quickly, which is of great advantage in the forest's frequent dampness, for dry leaves photosynthesise more efficiently than wet ones.

Another characteristic of rainforest plants' leaves is that many are brightly coloured when they first unfold and only later turn green. New shoots may be pink, brilliant red, pale yellow, rusty brown, golden, purple or variations of these colours. Some species of satinash can be especially brilliant. When there is a flush of new growth these trees make vivid splashes of rose-red in the forest. The red pigments contain a fungicide and also strongly absorb ultra-violet light and so protect the new shoots when they are most vulnerable.

At first glance the slender tall trees in a rainforest look very similar. The trunks are straight without branches below the canopy and have mostly thin, smooth bark. This superficial similarity, however, masks the tropical rainforest's single most characteristic feature: the great diversity of species of trees. Seldom do you find two individuals of the same kind growing side by side.

What, then, is a rainforest? It is a place with virtually no limits to plant growth'; where trees are so dominant that forests grow within forests; where competition for their energy source, the light, is so fierce that plants grow upon plants, that trees strangle each other and vines claw and push their way to the canopy where they smother their supports. In rainforest trees may have buttresses, flowers sprout from their trunks, leaves unfold in brilliant colours. Rainforest is not a wild chaotic place but one where trees, vines, shrubs and epiphytes, together with all the animals they support, form a self-sustaining, dynamic association of species, the richest, most diverse on earth.

Brightly coloured new foliage, like on this plum satinash, is a characteristic of tropical rainforest.

On the Trail of the Cassowary

Mission Beach, 13 February

I had heard of Joan Bentrupperbäumer. She is said to ride a motorbike cross-country to her study site. There she follows cassowaries on foot through trackless dense rainforest full of wait-a-whiles and mosquitoes. You might expect some large strapping person full of bravado.

But on this hot and sticky afternoon I meet a small, even petite woman, barefoot and looking cool in a white shirt and pink slacks. Joan has short dark hair. Her brown eyes are set in an expressive face. As we talk I discover that she has a passion for the rainforest and the giant birds she is studying. This passion, however, is firmly under control and she talks coolly about both forest and cassowary. She was once a radiologist in a hospital. She is the kind of person you would like to see there—calm and capable, yet not without imagination.

I had arranged to meet Joan because I hoped she would tell me something about the life of that huge and enigmatic bird, the cassowary.

We talk in her study, a comfortable room hung with posters and photographs of cassowaries. A large aerial photograph of Joan's study area leans against one wall. On the end of her workbench stands a pottery bowl filled with cassowary memorabilia: large pieces of eggshell of a surprising green colour, the skull of a bird that was killed on the road, feathers and nests of small birds that used cassowary feathers in their constructions. The windows look out onto a lush garden full of palms, ferns and giant cunjevois. Dotted among them are heliconias bearing scarlet flowers. Orchids and bromeliads drape the trees. Beyond are views of the beach and Dunk Island.

I ask Joan how she became involved with cassowaries.

'It started in February 1986,' she says, 'after cyclone Winifred, which caused amazing devastation to the forest. I'm from this area and I've been through quite a number of cyclones, but I'd not seen anything like it. There was complete defoliation over huge areas. The winds came at night and the next morning there was dead silence, absolute dead silence. The only thing I could hear was the occasional call of an owl. Which was unusual. It was like I imagined the forest would look like if some defoliant had been used; not a leaf was left on the trees. A lot had come down in winds that reached 180 kilometres per hour. There was very little rain.

'We didn't know how many cassowaries may have died, crushed by falling trees and things like that. We noticed a sudden appearance of the birds in the urban areas around Mission Beach. What was really strange was that for weeks after the cyclone there was still no rain. That, coupled with the loss of foliage, meant that the cassowaries were totally exposed to February's incredibly hot summer sun. The birds' black feathers absorbed a lot of heat, which caused them considerable stress. The result was that the cassowaries came out of what was left of the forest, entered people's gardens and sat under sprinklers to cool down.

'New leaves began to grow almost immediately but these were soon chewed up by caterpillars and other insects. It was a very hard time for all wildlife. The local people became very concerned. They were saying, "What's happening? I'm getting cassowaries in my backyard. They're starving to death. What are we going to do about it?" The last question was really aimed at the Queensland National Park Service. Their response was that perhaps a rescue operation should be mounted and a feeding program established. So that was done. Fruit for the birds came from as far away as Shepparton in Victoria; spoiled fruit mostly that was sent up by train. The

National Park Service set up feeding stations in the forest. That's when they asked me if I would be interested in trying to find out what the cassowary population was in the area, because we had just no idea. I said I was and that's how it all started.

'I found that about twenty birds were being fed, though this was not the entire cassowary population. More disturbingly, I was also recording quite a few deaths. After the cyclone, when the birds came out of the forest, some were killed by cars. Others, mostly chicks, were killed by dogs. Some died of disease and starvation.

Dillenia is a male. His casque is smaller, is jagged right across the top and curves over slightly. He has a red spot below his bill and his right wattle is split.

Jasminum, a female, has a jagged edge on the top of her casque. The purple at the back of her neck reaches right up to the level of her ear. One of her wattles has a swelling on it and the blue on her head extends to the front of her eye.

Kamala, also a female, has a high casque, smooth across the top and angling sharply down at the back. The purple at the back of her neck does not reach as high as her ear. One of her wattles has a piece out of it.

Superficially all cassowaries look alike. So when Joan Bentrupperbäumer had to know individual birds she learned to recognise all the subtle characteristics that make each animal unique.

'In 1990 I got a grant from the Australian National Parks and Wildlife Service, in association with the CSIRO, as part of their endangered species program, to study the birds full time. The cassowary is definitely an endangered species. Some estimates say that there are between three and four thousand left. But I think there are a lot fewer than that, perhaps as few as one thousand fully adult birds. In the area around here, which has the densest population of cassowaries, there are only sixty-three. In many places, on parts of the Atherton Tableland, for example, they have disappeared from some areas and are declining in others.'

So Joan began work to find out as much as she could about this bird, especially its requirements of food and space in the rainforest. In order to do that she must observe cassowaries closely for long periods. I ask her if she uses radio collars or similar devices.

'No, no!' Joan exclaims. 'I think you gain a lot more information about a population of animals if you don't traumatise them in any way. Cassowaries are easily traumatised through handling and also easily injured. Their beaks and their casques [the helmet-like growth on their heads] are quite soft and susceptible to damage. You often see birds with broken or twisted beaks in zoos. Cassowaries don't like to be handled at all.

'I think if you want to study an animal's behaviour, to observe its interactions with others of its kind and also with other species, you need to follow it for up to eight hours a day. You can't do that if you traumatise it, you need it to get used to you. I've now habituated four cassowaries but it takes a long time.

'To begin with I concentrated on one male. I had worked out the area he moved about in and then I used to be in that area as much as I could. It was important that I didn't hide myself or try to sneak up on him. I had to be visible, but also quiet—the birds are very sensitive to noise, especially footsteps. Human footsteps sound much like those of cassowaries. Whenever a bird heard me walking about, it would boom at me as it would at another cassowary entering its territory. Gradually I made longer and longer contact with that male—five minutes, then ten and so on. I did that for four months, and by the end of that time the bird was completely used to me. As far as he was concerned I was just part of the environment. I follow him now for eight, or even ten, hours a day. If I'd traumatised him and put a radio on him, I doubt whether I would have got so close to him. The information I've got from getting this bird to accept me has been fantastic.'

Rainforest fruit are the cassowary's main food. These crimson berries are eaten by catbirds and by king parrots and more than likely are also part of the cassowary's diet.

Before telling me about her adventures with this male and the two females he associated with, Joan explains that she had to learn to recognise individual cassowaries and also to tell a male from a female. This is not an easy task with cassowaries. It can only be achieved through constant contact with the birds over long periods of time. Joan looks at a bird's distinguishing marks and enters these on an identification chart, a kind of mug shot. First she looks at the bird's casque. She notes the helmet's shape and size, if it is tilted to one side and if it is scarred in any way. The two fleshy wattles that hang down from the front of the birds' necks can also bear distinguishing marks; they may be split or torn in some fight or one may be longer than the other. Joan also looks at the corner of the bird's beak and the colour pattern at the back of the neck. Armed with these mug shots, Joan can recognise all the cassowaries she sees regularly. Telling males from females is not so simple. Joan has found that the males' tail feathers are longer and that females are of a slightly heavier build and have larger feet.

Once you know a certain number of individuals it makes discussions about them easier if they have names. Joan has given each of her cassowaries the scientific name of one of their food plants. Among the males there are Calamus, Acmena, Ficus, Faradaya and Myristica, while Eugenia, Licuala, Helicia and Bowenia are females.

Having habituated a male cassowary, whom she named Dillenia after a handsome tree, knowing other individuals and being able to tell males from females—all things no one had done in the wild before—Joan unravelled more of the mysteries about this most mysterious of the rainforest's birds. Joan tells me the story of Dillenia and two females, Jasminum, who was the younger of the two, and Kamala.

'It began in May. I was following Dillenia quietly and keeping my distance; I had to as the females were not used to my being so close. One day Jasminum approached him. Dillenia began to dance and pace all around her, now and again bumping into her with his chest. This went on for fifteen minutes or more. Jasminum then walked off and he followed her. Eventually they mated. It was the female who had sought

The matchbox vine, which has flattened stems, is just one of many climbing plants that make the following of cassowaries in lowland rainforest so difficult.

out the male. But not long afterwards I saw Dillenia approach Kamala. He danced all around her too, and from their behaviour I am pretty sure they mated, though I did not actually see it.

'The two females seemed to be rivals; a few times I've seen them bluff-charging each other at places where I thought their territories met. Confrontations begin with this incredible booming. You hear it often in the forest and it means you have intruded on a bird's territory and you had better move off. These two were doing a lot of booming and then charging short distances, turning, pacing about and charging again. They never actually made contact. It was all a bluff. Once when I watched these two bluff-charging a third female, called Eugenia, joined in. That was an extraordinary and exciting thing to see, those three birds, each larger than I am, pacing and charging in the forest.

'Eventually the nest site was chosen. To begin with no nest was made and the eggs were laid directly on the leaf litter in the early part of July. Later, while sitting during incubation, the male pulled litter towards himself with his beak and so built up some kind of nest. Unfortunately I didn't see the actual laying. Perhaps I can observe that when I have habituated the females. But I didn't want to disturb them at this critical time. Three eggs were laid. The first two were a very dull green and the third was a bright green. The shell pattern of the third egg was also different. It's quite possible both females laid in the one nest, but not having seen the females lay I cannot definitely say that is the case.

'From then on only the male was at the nest. He did all the incubating by himself. Observations at this nest, and others I've watched, showed that he never leaves the nest during the entire incubation period, which lasts from forty-eight to fifty days. He does not eat or drink. He does get up now and again to roll the eggs.

'Two eggs hatched but not the third. Dillenia was now torn between incubating the remaining egg and taking the chicks out to forage, for they were peeping with hunger. Dillenia would sit on the egg for a little while, then get up and roll it and turn it. Once he raced off and chased a sub-adult cassowary that had wandered into his territory. Another time he charged me, the only time he has done that, for he's incredibly protective of the young. But it was all bluff. He stayed with that egg for two days. Finally he left it and set off with the chicks. I followed them for about five hours and then went back to the egg. I was just in time to see a goanna roll it along and knock it against a tree. The egg cracked open and the goanna ate the content. It was infertile. There are a lot of stories about feral pigs eating cassowary eggs and competing with the birds for food. From what I've seen that is not so. Cassowaries chase pigs off, not the other way around. That is down here, at Mission Beach, it may be different in other places.

'I continued following Dillenia and his two offspring, but I never attempted to touch the chicks. That was difficult in the beginning, when they were tiny. When they were out foraging in one place for some time I'd sit down on a log. Often they came up to me and would peck at my ear-rings or pen or whatever took their fancy. I had to be extremely careful not to move suddenly or there would be an awful kerfuffle and they'd rush off. They could easily lose their confidence in me that way. I also had to be careful they would not follow me as I walked away. To them I was just like their father, two legs moving off. They followed me once for a short distance. I thought the male might take exception to this but he was unconcerned and he didn't do anything. One of the chicks died, but the other is now a sub-adult and is still used to me. I called him Dillenia alata [this is the full name of the tree the adult male was named after].'

Joan has become deeply involved with these giant birds. How does she feel about them, and particularly about being so close to them out in the forest, entirely on the birds' terms?

'I find it a great privilege to be able to work with a bird like that. I spend a lot of time with them and they've become unbelievably tolerant of my presence. That to me is really fascinating, and has enabled me to learn so much about them. I've learned how to behave towards the birds, that there is a limit as to how close you should go. I can now also recognise a bird's intention if he wants to feed where I happen to be standing. My feelings for the cassowary are hard to explain. It's something very special that only people who have worked at such close quarters with wild animals understand. I haven't met many people like that. I've become so involved with the birds and so interested in their behaviour, habitat use, vocal communications, mate selection and many other things, that I want to study them for a long time yet. So far I've just skimmed the surface.'

Joan was born in Tully, not far from Mission Beach, 40 years ago, so she has known rainforest from earliest childhood. It holds no terrors for her. Even as a child she was out hunting pigs with friends. In those early days she admits it may well have been a show of bravado. But this matured into a genuine love of the forest. Joan explains further.

'I would find it very difficult to live in a dry area now. People who have done studies in the rainforest often tell me they find it oppressive and difficult to work in for long periods of time. I don't have that problem at all. I think that is

Thick lianes are characteristic of tropical rainforest.

due to my association with the forest when I was young. That must have been when I developed my sense of direction, of not being afraid of getting lost. It is important, I think, this feeling that you can always find your way out and also that the place does not harbour anything frightening at all. To me, the rainforest is much safer than any urban area.

'Even as a small child I found the rainforest very attractive. I enjoyed a sense of freedom there, a sense of being alone with it, I guess, of being at home in this immense forest that not many people would go into. It was a feeling I kept to myself because I knew it was not something that was understood. I couldn't share it. Most people find the forest quite threatening. Even the locals usually think of the rainforest as something to be conquered or destroyed, also it harbours vermin, they say. I used to know a few pig shooters who really liked to go into the forest. But that's more a macho thing I think; conquering again through killing its animals.

'It's only in the last few years that I've been able to get back to the forest. Before that I worked in hospitals. I spent some time in England and studied there. Then I did a number of years of volunteer service in Nepal in radiography and in clinics treating tuberculosis. When I came back to Australia, and this place in particular, I felt I wanted a job I enjoyed doing. I knew hospitals weren't really the places I wanted to be. So that's when I decided to do a degree in zoology. Now it's just marvellous to be able to work full time in the rainforest with these amazing birds.

'It can be tough in the wet season,' Joan adds. 'The rain is no real problem and I don't mind the heat. I just keep going. The cassowaries keep foraging in the pouring rain and I keep following them. During the wet I have to do a lot of survey work where I walk through places to see what cassowaries are present. I may not actually see the birds, but I can tell their presence by droppings and footprints. The places I have to go are full of mosquitoes, there are so many that any exposed skin, like your face or hands, is black with them. There's no way you can stop and sit down to have a break. I can't use a net around my hat or anything like that because these cyclone-damaged forests are full of *Calamus* [wait-a-while], which would soon tear any net. All you can do is plaster yourself with repellent but after a couple of hours your skin starts to burn. I can't last any longer than two hours. So every two hours or so I go home for a while, write up my data, and then go back for another stint. That way I won't get so depressed about it that I don't ever want to go back. Luckily this difficult time lasts for only a few months.'

Today has been such a hot and humid day. But it is getting dark and cool now. The beach below the house and the islands beyond look inviting and benign. The garden Joan and her husband have planted has few mosquitoes. Sunbirds flit from flower to flower. Yellow-spotted honeyeaters call from the bushes. As I take my leave a large-tailed nightjar hammers out its call on the forested hill behind the house.

Along the River

Mossman Gorge, 18 February

The jungle perch is a predatory fish of lowland rainforest streams.

Apart from small gudgeons and large eels, both species that can climb waterfalls, there are no fish in Bulurru's pools. To see more of the fish of the rainforest streams, I travel north to the Mossman River. Conditions are ideal. The monsoon finally retreated a few days ago and this morning it is sunny after a showery night. There has been sufficient rain to make the river flow strongly, but not so much that it is raging and foaming with its waters less than perfectly clear. Rising as it does in rainforest, the Mossman River never becomes muddy. It flows among granite rocks of all sizes, from as large as a house to as small as a pebble.

I stand on one of the huge blocks of stone and watch the water boil over some boulders then flow into a deep pool, its waters faintly tinged with green. Jumping from rock to rock I make my way to the pool and immediately find what I am looking for. About forty jungle perch cruise the waterhole. With casual sinuous movements of their fins and tails they remain almost motionless in the strong current. Some are tiny fingerlings but others are large broad-headed fish 30 and 40 centimetres long. Their backs are greeny-bronze, their sides silver speckled with darker colours.

As always at this time of year on the lowlands, the day rapidly loses its fresh coolness. Already it is hot and humid. I cannot resist the temptation to slip into the pool with the fish. At first they retreat behind boulders at the opposite end, but soon, one after another, they swim inquisitively towards me. Some of the smaller ones nibble the hairs on my arms. A large

Rainbow fish feed on insects that fall in the water but also on the eggs and larvae of other fish.

one brushes against my leg. I stand quietly, up to my chest in water and after only a few minutes they regard me as just another boulder. Every so often a dead leaf, a bark flake, a spent flower, a piece of lichen or other debris from the forest gently drifts down to the pool. As soon as anything lands on the water surface two or three jungle perch dash towards it with a speed too quick for the eye to follow. The debris is passed up. But when a small fig falls from a tree over the pool, it is quickly eaten. Any insects that land or fall in the water—ants and marchflies seem to do so more often than others—instantly disappear in a whirlpool made by a striking fish.

Dragonflies skim low over the water a little downstream. I am entranced by their colours and aerobatics as they hunt small flying insects or court each other. Suddenly a jungle perch leaps clear of the water and snaps up one of the fast-flying insects.

At first it looks as if a large snake squirms in and out among the rocks beneath the water. Then I see its head and realise it is an eel, as thick as my wrist, probing the cavities between and under the rocks. A few small jungle perch shadow it closely, ready to pounce on any small shrimps, dragonfly nymphs or other life the big fish may stir up.

Jungle perch are aggressive predatory fish and no others share this particular pool with them. Even the eel soon moves on.

I make my way back to the bank and walk upstream for a short distance. Climbing over boulders and squeezing between others I find a small pool in the back eddies. It is close to the bank and in heavy shade. Wearing a pair of goggles and a snorkel, I sit down in the quiet water enclosed on three sides by huge boulders.

Underwater, most movements are slow and fluid. Tadpoles wriggle along the sandy bottom. At the entrance to a small cavern a freshwater prawn feeds on green algae. A crab scuttles between pebbles at the shallow end of the pool. Gradually I become aware of small fish all around me. Most conspicuous is a school of a dozen rainbow fish. Some are brightly coloured with red on their wide fins and tails. They are very inquisitive and swim right up to my goggles, even nibbling the glass with their tiny mouths. At the bottom of the pool purple-spotted gudgeons with blue and purple flecks on their sides, cling to the rocks. Four small fish, appropriately called blue-eyes, pass slowly, elegantly, a metre or so in front of me. Their fins are yellow edged with pale blue, as are their almost transparent tails. Other fish, small, translucent and almost the same colour as the rocks, swim among the boulders.

I lie on my back for a moment, underwater, and look up at the rainforest. Some rainbow fish swim above me. The water is so clear that for a moment it seems the fish are moving through the leaves. Now and again a fish swims up to gulp some air or a leaf falls, disturbing the water's glass-like surface. The trees and shrubs then shimmer and sway back and forth as if they were made of rubber. It is an utterly peaceful and tranquil world where all movement is graceful.

A March fly falls into the pool. Instantly the languid mood is shattered. Five or six rainbow fish rush the unfortunate insect and tear it to pieces.

North of the Daintree

Cape Tribulation National Park, 23 February

The lowlands north of the Daintree River, their forests and landscapes, are very different from the uplands of the Tableland. The most noticeable difference at this time of year is the constant energy-sapping combination of heat and high humidity. Whenever you stop mosquitoes crowd onto every bit of exposed skin. These are mere discomforts, however, which I soon forget for I have arrived at one of those magical places that encapsulates a whole region's wonder and quintessence.

I stand halfway up a ridge. A rivulet so steep it is almost a waterfall, trickles and whispers through a forest of tall trees. Fan palms with gigantic long-handled circular leaves grow out of a lush undergrowth. The leaves of shrubs, ferns and vines range from the large to enormous. King fern fronds four metres long arch over me. These are, in fact, the world's largest ferns. I am also dwarfed by ginger plants as tall as the native bananas among which they grow. Some small trees and shrubs have simple leaves the size of a sheet of writing paper, while others have compound leaves a metre and a half long. As always the vines create the feeling that all this growth is out of control and is about to overwhelm anything and everything that stands still. Climbing aroids wrap tree trunks in fresh new leaves. Where a cyclone

W.T.Cooper 92

Where the rainforest meets the beach the satin touriga or beauty leaf trees lean so far out towards the light that their trunks are almost horizontal.

punched a hole in the forest a few years ago, re-grown trees are smothered by a profusion of heart-shaped leaves of the vine *Merremia peltata*. Large leaves are characteristic of lowland rainforest. As you move to higher and higher elevations, the leaves become smaller and smaller. On the mountain tops most trees have tiny leaves only a few centimetres long.

The same cyclone-created gap gives me views north along the coast. The bulk of Thornton Peak towers above me. Its upper slopes are wind-swept and lead to a 1374-metre summit of pyramids and towers of granite. Only the very top is hidden by a small white cloud. It is much more usual, at this time of year especially, for the mountains to be wrapped in cloud and rain. Steep-sided ridges reach out from the peak (I am standing on one of these) and then bend towards the sea where they end abruptly in rocky headlands. In the embrace of the mountain's arms, between the headlands, yellow sandy beaches press against the rainforest. Some are several kilometres long, others are small intimate coves. Streams draining the peak and the ridges cut through the beach to the calm blue sea.

A slight movement on a tree right beside me catches my eye. Slowly I turn my head and see something quickly move out of view. Even more slowly and carefully I walk the few paces to the tree and peer around the trunk. A red-brown eye surrounded by scaly lids peers equally cautiously back at me. As I move further around, so does the lizard. I keep moving and eventually the rainforest dragon seems to tire of dodging round the tree. It stays put, its sharp claws digging into the tree's bark. This lizard, its head and body about 15 centimetres long, gets its name from its crest, topped with sharp, triangular, tooth-like scales. A row of similar scales runs down its back. Despite its fearsome name and appearance it is not at all aggressive. I bend down till we are eyeball to eyeball. I can see the polish on its claws and the pale inner-ring of the pupil of its eye. An ant runs over its head. The dragon wipes it away with its front foot. Staring at it so closely, I have also become an irritation. The dragon inflates his yellow throat pouch then, quick as a flash, runs up the tree.

A few kilometres south of Cape Tribulation I stop on the banks of Oliver Creek, at an expanse of very rare and special rainforest. It is rare in the sense that all lowland rainforest is rare. They were the first to be cleared to make way mostly for sugar cane growing and later cattle grazing. Only a few pockets of this kind of forest remain and most of them are here in the Daintree–Cape Tribulation area. It is the only place where you can see tropical rainforest growing from high mountain peaks to the beach in an unbroken sweep. The forests at Oliver Creek and nearby Noah Creek are important for two other reasons as well. Firstly it is one of the very few areas where primitive and ancient flowering plants are found right at the coast. Secondly, it was at these two creeks that the far-reaching significance of these plants was first discovered. That was in the 1950s and '60s. It was a discovery which led to a revolution in our thinking about this tropical rainforest and its long evolutionary history on this continent. This new thinking in turn marked the beginning of a protracted conservation campaign that eventually resulted in all of Australia's publicly owned tropical

Rainforest dragon.

rainforest being conserved through World Heritage listing. The Oliver and Noah creek area is the cradle of all the new thinking and new attitudes to Australia's tropical rainforest. We are all the richer for it.

To me the ancient lineage of Australia's rainforest and all that it means and implies is the single most exciting and stimulating subject in the wet tropics. It is a theme I will return to many times, in visits to Mount Lewis and to Boonjie and in discussions with Geoff Tracey and Geoff Monteith.

After all the wet season rain the Oliver Creek forest is marshy. I would be over my ankles in water and mud if the National Park Service had not built a raised boardwalk. While in these cyclone ravaged lowlands trees rarely reach truly gigantic proportions, my view is still of column after column of dark, straight, and large tree trunks rising out of a froth of fresh new green. The early sun slants through the trees, highlighting this plant or that in its spotlight. Palms are especially numerous, lending a great elegance to the forest. Low down grow the lightweight walking stick palms, less than two metres tall, draped with long strings of yellow and red fruits. The sturdy, feather-leaved black palms are well suited to these waterlogged soils; their stilt roots lift their trunks above the water. A shaft of sunlight strikes one of these palms' large bunches of deep pink fruit, a splash of colour in the general greenness. And, of course, the wait-a-whiles are inescapable. Backlit, the dagger-like golden spikes on the yellow wait-a-while stems look threatening and make a striking contrast to the plant's soft, down-curved leaves.

On the marshy lowland soils black palms grow on stilt roots.

Compared to the Tableland, the lowlands are strangely quiet. Birds are few. I am, therefore, immediately drawn to flocks of ten and twenty black birds speeding overhead and to the chatter of many more coming from a tall tree. The focus of the hundreds of shining starlings is an emergent briar silky oak. The tree's outer branches are weighed down by clusters of nests made of vine tendrils and strips of grass. There are well over 400 of the bulky domed nests. Most of them contain well-grown young. Whenever a parent bird lands at its nest two or three gaping mouths are thrust towards it, begging for food. Insects form a major part of the nestlings' diet but they are also fed a certain amount of fruit. For this the adults do not have

Shining starlings nest in colonies in emergent rainforest trees.

far to go. About twenty metres from the nest tree grows a dark-leafed fig laden with small orange fruit. The fig is a much smaller tree and by standing underneath it I can get quite close to the starlings. I can clearly see their glossy iridescent plumage. But what really stands out are their brilliant scarlet, almost incandescent eyes, which give them a wild and fierce look. There is a constant shuttle of birds between the fig and the silky oak. This same species of starling comes to Bulurru to feed on the celerywood and other fruiting trees. But they do not nest at the higher altitudes of the Tablelands. In another month or so nesting will be complete and the starlings will make their way back to New Guinea.

Birds may be few but these forests do not lack colour or animation. It is provided mostly by the butterflies. The ubiquitous and splendid birdwings and Ulysses dance in and out of the sun high up in the forest. Lower down a red lacewing flutters around a vine that bears its name, so called because the butterfly's caterpillars eat its leaves. Orange cruisers and brown and yellow lurchers fly lower down. Triangles, blues, yellows, jezebels, tigers and others wink and flash their colours throughout the forest.

Cape Tribulation is formed by an arm of another peak, Mount Sorrow, plunging into the sea. To the north arcs a beach, a low sandy rise, that meets the rainforest in a sharply demarcated straight line. The trees' roots shy away from the salty sand at the beach front. I can actually see where some of the roots turned back at the high water mark. Others, especially those of the buttressed satin touriga or beauty leaf, make a right-angled turn and then run parallel to the beach, forming a low wall many metres long. Because of this saltwater barrier, the trees' roots have to be anchored in the salt-free soil above high water mark. But their trunks, defying these limitations, lean out over the beach in their competition for the light. Again the satin touriga does so spectacularly. Their trunks, as large as any in the forest, are almost horizontal. To support this massive weight they put down branches to the sand, making them look as though they are leaning on their elbows. The whole forest appears to be tilted 90°; the trunks growing horizontally, their crowns forming a solid wall at the waterfront, complete with vines. Walking along the forest edge I certainly feel that I am at the canopy. Fruits and flowers are low down, right to the ground, as are the butterflies that feed on them. I can just reach up and pick one of the round red fruits off a vine. They are sweet, but the flesh sticks fast to the seed and no matter how long you chew and suck on it you can

A red lacewing butterfly and the lacewing vine in fruit. The plant is the food for the butterfly's caterpillars.

never really eat the flesh. It is a bit like an all-day sucker and was given the name lolly berry as a result.

Many of the beach-front trees have beautiful shiny pale green leaves. This polish may be a protection against the constant salt-laden winds. One of the most handsome trees in all the tropics, the red beech, also grows along the shore. Its red-brown bark flakes off in fine layers. Its crown, bending down to the ground, allows me to admire what look like two kinds of flowers. One kind has yellow petals with a red centre, the other is like a red star with a white centre. But when I look closer I soon realise that the red 'flower' is really an opened fruit holding seeds covered in white arils.

Male shining flycatchers berate each other in a territorial dispute.

In a few places the low sandy ridge, too low to be called a dune, holds back ponds of fresh water seeping down from the hills. The waters are dark with tannin from fallen leaves and other debris. The vegetation around them is dense. These places seem very mysterious. They harbour countless mosquitoes. Long ago they were also the haunts of saltwater crocodiles. Chest-high crinum lilies, ringing one such pond, have large heads of white flowers. Beside them a tree with dark green foliage also carries bunches of white flowers. Many of them have matured into glossy burgundy coloured fruits the size of ducks' eggs. These grey milkwood fruit are curiously light-weight and many float in the pond. Perhaps these seeds are dispersed by the tides and flooding creeks.

At one point a wide swathe of mangrove trees, growing in rocks and mud, invades the sand. From among the stilt roots a beach stone-curlew regards me steadily through huge yellow eyes. Where the mangroves are closest to the forest, vines have reached out and fastened on to them. Sturdiest of them is the matchbox bean. Its still-green pods, over a metre long, are strange decorations for the mangroves. When ripe the pods will contain flattened round seeds, very hard and durable and a dark brown in colour.

They look like slices of fine polished wood about five centimetres in diameter. In the old days forest workers used to make containers for their wax matches out of them.

From the depths of the mangrove patch come the scratchy scolding notes of small birds. It is the kind of agitated sound they make when they spot a snake or other predator. The two glossy, small black birds are not scolding a possible attacker but each other, perhaps in some territorial dispute. Both shining flycatchers are males (the females are brown and white) and face each other only a few centimetres apart. With crests raised, tails fanned, and showing off their orange-yellow gapes, they berate each other. I leave them to their quarrel and walk towards Cape Tribulation itself.

Sand gives way to rock. I walk to the summit, to the furthest point out into the sea. From this elevation I look north along the beach. Coconut palms project out of the tilted forest. To the west and south are Mounts Sorrow and Hemmant and Thornton Peak. I know they are there, but I cannot see them. Clouds have gathered around their summits and slowly roll down the ridges. The sun goes in. Soon it will rain again, a condition that has nurtured special forests on lowland and mountain top alike uninterruptedly for more than a hundred million years.

Peripatus

Bulurru, 3 March

The monsoon has returned. At dawn the low heavy clouds that sluggishly crawl in from the southeast grow almost imperceptibly lighter till they are a uniform mid-grey. Sunrise can only be imagined. Mist lies motionless in the valley below the house. For the first time in many days it does not actually rain but the forest is saturated. Water from last night's downpour still trickles down the tree trunks and drips from every leaf, every piece of lichen and moss.

Peripatus or velvet worm.

The mosses especially thrive in these conditions. They cover fallen trees with small pillows and soft blankets of vivid green that seem to glow in this semi-dark.

This all embracing dampness and virtual twilight are usually described as murky, sombre, gloomy, cheerless or dull—words that are also used to describe the darker, melancholy side of human moods.

It is usually during these days of endless rain that rumours of strange animals, monsters even, surface. I have been told of panthers (always black panthers), monkeys, giant bats, marsupial tigers and strange snakes having been seen in these forests. Reports of UFOs and mysterious lights are not infrequent.

I too can conjure up all kinds of interesting and exciting animals. My thoughts usually turn to some undescribed and highly colourful bowerbird, a never before seen exquisite possum, or a snake new to science. Today in the dripping dark forest with wisps of mist trailing across the paths I can imagine all kinds of undiscovered exotica. Cold reflection, however, soon dispels these fancies. It is just about impossible that there would still be any larger animals, among the birds and mammals especially, that have not been recorded and described. A new frog or reptile then? Perhaps, but unlikely. Undescribed invertebrates? Most certainly. There are many undescribed insects, especially among the beetles and moths. Most are tiny, nondescript species. It is among the minute animals in the brown dampness of the leaf litter that I find something extraordinary and, on a small scale, quite spectacular.

Bending down to pull yet another leech from my leg, I see a small, strange creature moving over the fallen leaves. It walks on 15 pairs of fat purple legs in a fluid rhythmic motion. On its head it carries two antennae. Its skin looks like velvet. This 'creature', barely six centimetres long, does not fit into any classification, any animal group I can think of. It is soft, like a worm, without the segmented body of an insect or spider. It is not like a centipede or millipede which, like the insects and spiders, has an exoskeleton, that is a hard outer skin. But its antennae and legs mean it does not belong with the segmented worms, which include the earthworms and leeches, homing in on me from all directions. However, the creature's body is quite wormlike. What is it? The missing link that connects the segmented worms with the arthropods; the insects, spiders, centipedes, crustaceans and the like? It is something stranger than any imagined monster.

For many minutes I dig deeply into my memory of early zoology lessons. Something vaguely rings a bell. Suddenly I have it: it is a peripatus or velvet worm. It *is* thought to be that missing link. Peripatuses have been classified into a group of their own and they are found in most of the world's wet tropical zones. They are a very ancient group going back to the Cambrian period, 570 million years ago, when, most remarkably, they lived in the sea.

Having no hard outside covering, a peripatus easily dehydrates. It must, therefore, stay in the dampness of rotten logs, among rocks or in ground litter. Only on these days of saturated humidity do they come out in the open. It is then that they live up to the name peripatus, which is derived from peripatetic, meaning travelling or wandering. I bend down to touch this stranger, to feel its velvet texture. The moment it feels my finger lightly on its back, it shoots out long strands of white sticky slime from its mouth. This has no effect on me, but a small predator, such as a spider, could easily become enmeshed in, or at least distracted by, these threads of glue.

Two Australian zoologists have studied this continent's species of peripatus for several years. Before their study began about eight Australian species were known. They added 40 more. They discovered that the male peripatus produces parcels of sperm from a depression between his antennae. But they never learned how this package is presented to the female. The females of some species produce live young while others lay eggs that take as long as 17 months to hatch.

To other small animals peripatus is a formidable predator. The same sticky threads that enmesh and deter attackers can also be fired with great accuracy at small insects. These become entangled and are then swiftly caught and dismembered by the peripatus's sharp cutting jaws.

There *are* 'monsters' and strange animals in these forests, stranger than any science fiction writer could imagine. Most of them are on a small scale and rarely reveal themselves to inquisitive humans. But if you look closely, have patience and perseverance, you will find hundreds of such miniature beings, each with its own intriguing story.

Rhinoceros beetle. Its larvae feed in damp decaying accumulations of leaf litter.

Fungi are the principal agents of decay in the breakdown of the forest's litter.

Recycling

BULURRU, 6 MARCH

The monsoon departed on one of its periodic retreats yesterday. Day break is cloudless, fresh and clear. Birds sing with renewed vitality. There is a glad-to-be-alive quality to the air. Animals come out of hiding and seem to rejoice. Perhaps my observations are coloured by my own good feelings and the birds, reptiles and insects are merely busy making a living after the inhibiting rain. But the bridled honeyeater bathing by flying into dew-laden leaves appears to do so with joy, the brown pigeon rising steeply into the air then gliding down seems to find enjoyment in its aerobatics, the black-faced flycatcher singing outside my window to my ears has greater energy than usual, and the male brush-turkey, his yellow wattle almost touching the ground, stretches his neck and beats his wings while standing on tiptoe with all the appearance of sheer exuberance.

As the sun rises the vegetation dries off. On the ground, inside the forest, dampness still reigns and will continue to do so for many days. But the increased light, the pools and dapples of sun, the backlit new leaves, some pink, some tender pale green, the fallen fruit glowing among the brown carpet of dead leaves, spread an aura of freshness and vivacity here as well. Small skinks, shining and iridescent, bask in the sun or chase after ground insects. In this atmosphere of warmth and humidity you imagine you can see the plants grow and feel the energy abroad.

The leaves, twigs, branches and even whole trees that constantly come down in the forest accumulate on the ground. This debris is no longer saturated as it was a few days ago but remains soft and damp, retaining moisture like a sponge. There is a not unpleasant smell of decay, not the stench of large organisms rotting away, but a faint sweet and damp scent; a whiff of the second of nature's energy-producing processes at work. The first, photosynthesis, is driven by chlorophyll, the green stuff of plants, reacting to sunlight. The second process is almost entirely carried out by plants and animals. Many of them are minuscule and the vast majority microscopic. Their work is performed hidden in the layer of dead leaves and other organic matter lying on the ground and the top 10 centimetres or so of the soil. This process of decaying the organic material, breaking it down to its component parts, which are then once again taken up by the plants as nutrients, is called the litter cycle. It takes place in just about all terrestrial ecosystems, but in the constantly moist and warm conditions of tropical rainforest it goes on at breathtaking speed and with great efficiency.

In Bulurru's forests eight to 10 tonnes of litter falls on each hectare every year. You would imagine that this amount would produce mountainous heaps of leaves, branches and logs. But this is not the case. Under normal circumstances only a thin layer of dead leaves covers the soil and only the most recently fallen branches and tree trunks lie on the ground. The exception is the immediate aftermath of a cyclone, when the ground may be strewn with an impenetrable mass of tangled vines and fallen trees. Even these accumulations, however, will rot away in four to five years. There are exceptions. The grey satinash, for example, may resist the attacks of borers, fungi and other organisms for decades. In drier habitats, such as a eucalypt forest, the story is very different. Fallen trunks may lie in the undergrowth for a century or more, and if it were not for periodic fires the litter build-up would be enormous.

As I walk slowly through a patch of sheltered tall rainforest, I see the evidence of the litter cycle in action all around me. Old logs and branches are covered by fungi; outgrowths of orange barred with yellow, of black edged with red, of plain grey or brown. One large stump is covered by a forest of delicate white fungi in the shape of umbrellas. When I try to roll a log it comes apart in my hands, its insides bored, tunnelled and excavated by insects, mostly beetles. Pill millipedes, opportunistic invaders of the insect tunnels, lie exposed and roll themselves into tight, shining black balls the size of marbles. Centipedes scuttle off into the leaf litter. Some of

In terms of numbers of species, beetles are the most numerous insects in the rainforest. They rival the butterflies in colour. The larvae of many species, such as the large longicorn on the left and the blue and red jewel beetle, live in dead wood which they help to break down. From left to right: longicorn beetle of the sub-family Prioninae, the green Anoplognathus smaragdinus, *the jewel beetle* Stigmodera rollei *and* Rhipidocerus australasiae, *a longicorn with fan-shaped antennae.*

the tunnels contain the pupae of stag beetles and longicorns, large spectacular species, but most of the inhabitants are tiny. I roll another log. It too disintegrates. One of the larger tunnels is the home not of an insect but a lizard, a spiny skink. It lives only in and under damp logs where it feeds on insects and other small animal life.

The rotting wood is permeated by fine networks of white threads. These are the growing parts of the fungi called mycelia and are the major agents of decay. Mycelia spread rapidly through damp, softened wood and also dead leaves. What we call the mushroom or toadstool is only a very small part of a fungus. It is the reproductive body that carries the spores, the equivalent of seeds in green plants. The growing mycelia can penetrate solid wood and consume it, but only slowly, a cell at a time. So the more wood is exposed the more rapid the process. Boring and tunnelling insects speed up the process, not only by opening up the wood but also by carrying spores into their tunnels. Animals that hunt woodboring insects accelerate decay even further. Many of the logs around me have been pulled apart, reduced to chips, by striped possums and white-tailed rats in their search for beetles and cockroaches. It will only take a few months for the chips to be reduced to humus.

It is not just fungi and woodborers at work here. As I carefully peel back a layer of dead leaves from the damp soil I expose a mass of jumping and wriggling minute animals. Some I can make out with the help of my reading glasses. These are the larger springtails, which are insects, and amphipods, which are crustaceans. I also see small cockroaches, beetles, ants, snails, slugs and the end of an earthworm disappearing into the soil. Others are so small that even a strong magnifying glass is of little help, but a microscope would reveal a world seething with small springtails, mites and nematodes. In north Queensland's rainforests very little is as yet known about these minute animals; most can only be assigned to broad categories. But the variety of species is enormous. In an English forest,

The spiny skink lives in damp hollow logs where it feeds mostly on insects and other invertebrates that help to recycle the forest debris.

which is much less diverse than tropical rainforest, 1000 animal *species* were found in a square metre of litter and soil.

Not only are there many species in the litter, most occur in numbers that defy the imagination. The weight of living animals in a forest may perhaps give a little insight into just how busy a place the litter zone is. In an Amazonian rainforest (no figures are available for Australia), all the living plants on a hectare of land weighed 1000 tonnes. All the animals, from jaguars and monkeys, birds and snakes, down to insects and soil organisms, came to .21 tonnes or 210 kilograms per hectare. Soil and litter animals accounted for 75 per cent of this weight. When you realise that just an average sized bird weighs countless times more than a minute litter decomposer, you can gauge, if only a little, what great numbers and energy are constantly at work breaking down the fallen plant matter.

Just as the striped possum and the white-tailed rat help in returning logs to the soil, so birds like the chowchilla and the scrubfowl, with their constant scratching, help turn over the litter and hasten its breakdown; and that despite the fact that the birds eat a considerable number of the decomposers in the process.

The last stages of this breakdown take place in the soil. The vital link here is not a microscopic animal but the earthworm, a veritable giant in this world. The mass of partly decayed particles of leaves and wood are consumed by the earthworms and then worked into the soil. So thorough are the worms that nearly all the soil to the depth of five centimetres or more has passed through the alimentary tract of an earthworm at some stage. Once in the soil the final breakdown is effected by microbes, the bacteria, fungi and algae. By this time the forest litter has been reduced to humus, a dark brown amorphous material without a trace of the structure and composition of its original

components. All the nutrients that originally built the leaves and wood are returned to the soil.

This food is immediately taken up by the roots of the living plants. Even the largest trees have feeding roots close to the soil surface, and often in the litter itself. Rainforests recycle themselves endlessly and efficiently. So efficiently that next to no minerals are leached out of the topsoil and litter despite periods of prolonged and heavy rainfall. Streams sampled in mature rainforest contained little or no minerals derived from the soil. The litter cycle makes it possible for rainforest to grow to a climax community even on poor soil, in some places even on pure sand which itself does not contain enough nutrients to support such a forest.

Photosynthesis and the litter cycle working in tandem are the engines that drive tropical rainforest to be the most efficient and prolific of land habitats.

Big Feet

Chowchilla, 19 March

While walking at Chowchilla in the drizzling rain, Bill, Wendy and I spot a small bird, about the size of a quail, running across the track. It does not go far and huddles under some dripping plants. It is brown, has pale outsize feet and is covered in down except for its wings, which are fully feathered. The bird seems to be unafraid and we go close for a better look. Being surrounded by three giants, however, is too much for it and it suddenly takes fright and flies off. Just for a moment we are stumped as to what kind of bird this could be. We realise it is very young, but if so, how could it possibly fly? And where are its parents? The answer clicks into

Dominant male brush-turkeys chase all others of their kind, even females, away from their mounds.

place for all three of us at about the same time, prompted by the clucking and scratching of a brush-turkey on his mound just behind us. The small bird was a newly hatched turkey chick, which possibly dug itself out of the mound only hours before.

The male scratching away is one of several brush-turkeys that live around the Coopers' house. Between them these birds, by their constant raking and scratching, try to re-arrange the garden from time to time. This one is the dominant male. His virtually featherless head and neck are bright red and the equally bare wattle, which he wears like a fleshy necklace, is yellow. At this moment his wattle is drawn up, out of the way of his raking feet. When he struts the forest and clearings, trying to intimidate the other turkeys, the wattle hangs down almost to the ground.

Being used to people, the turkey ignores us. Standing in the crater-like depression at the top of his metre high mound, he digs a conical hole at a 45° angle. When he considers this deep enough, he thrusts his head and neck, right up to the yellow wattle, into it. He shakes and fluffs his feathers for a while then withdraws his head and fills the hole in again. Twice more he goes through the same procedures. He is taking the temperature of the mound with his sensitive tongue. The mound's temperature is critical for it is the incubator for the female's eggs. The male can regulate the temperature and moisture content of the mound quite precisely.

We walk on to the house, closely followed by the turkey. Bill scatters some food for the birds across a small patch of lawn. While the turkey feeds he lets his wattle hang down and cocks his fan-like tail. Both are signs of dominance. Whenever a female or another male appears he runs at them aggressively at full speed, his tail still cocked. The objects of his ire run from him with their tails and heads lowered in submission. It is understandable that the male turkey chases other males from the area around his mound, but his unrelenting aggression towards the females seems inexplicable. After all, his considerable efforts in building the mound and maintaining its temperature over a six month period are solely directed at providing an incubator for their eggs.

The male, strutting about and almost tripping over himself with red-in-the-face self-importance, at times cuts a comical figure. But he and his mate are among the rainforest's most remarkable animals. The brush-turkey and the scrubfowl, which also occurs in these jungles, do not build nests, do not themselves incubate their eggs, and do not in any way nurture the young once these have hatched.

Brush-turkeys and scrubfowl belong to a group of birds known as megapodes, which means large feet. With these feet they rake together mounds of ground litter: leaves, sticks, bark, mosses, lichens and some soil. Once it is heaped together the mound's material is broken down by microbes, mostly fungi, and other organisms. This breakdown generates the heat that incubates the eggs. In effect the mound is a large compost heap.

The brush-turkey mound at Chowchilla was begun in December after the season's first substantial rain. It came to the Coopers' notice when they saw a section of the path swept clean of all debris. For many days they

The scrubfowl, also mound builders, live in closely bonded pairs.

watched the male as he raked and raked and raked. Scratching mechanically, almost absent-mindedly, the turkey would rake two, three times with one foot then with the other, for hours on end.

It took several weeks for the mound to take shape, to settle and then to stabilise. The female had no part in the mound building. She too scratched, but only to turn over the litter and to pick up as much of the insect and other animal life as she could. She had to put on condition to produce the large and numerous eggs.

When the female is ready to lay she goes on to the mound, digs a hole about 60 centimetres deep, lays the egg in it and covers it. She must give the male some kind of signal when she is about to produce an egg, so that he does not chase her off. The female can lay an egg every three days over a period of five to seven months. If food is plentiful she can lay as many as 50 eggs in a season. The chicks take nearly 50 days to hatch, 20 days longer than those of other birds of the same size. Over the breeding season the first eggs hatch well before the last one is laid. It has been found that a mound can accommodate about 16 eggs at any one time.

For incubating eggs and producing young, the megapodes have a system of unrivalled efficiency. Also the mounds are raided by few natural enemies; large goannas are probably the only ones. Birds that raise their broods in the conventional way lose 80 per cent or more of their eggs and young before they are independent. Producing several broods in succession, the most young they could produce would be fewer than ten. A brush-turkey, by contrast, can send 40 or more offspring into the rainforest each season. But there is always a trade-off that balances things out. Attending the mound and producing the eggs are full-time occupations which leave the parents no time to look after the young. These solitary, inexperienced little birds are vulnerable to predators, to the elements as they are not brooded, and in certain seasons to starvation.

The birds' breeding success depends on two other factors: the constant temperature of the mound and the special adaptations of the egg and its development.

To keep a heap of dead leaves and sticks at a constant 33°C for up to seven months may seem an almost impossible task, but it is not as difficult as it appears. At the depth the eggs are deposited the mound's temperature varies little, even if it were left unattended. When the mound becomes too hot the male turkey opens it up and if it is too cool he covers it over. A covering of fresh debris only a few centimetres deep effectively reduces heat loss.

The megapode egg is large and elongated. The size of the egg and the long incubation mean that the chick is well advanced when it hatches. It has wing feathers and is, therefore, able to fly and has enough down to keep warm even if the temperature drops below 10°C. The egg's shell is very thin, which makes it easier for the young to break out of it. The chick's lungs are especially adapted so that it can breathe down in the mound. Once it has struggled free of the egg the chick's problems have only just begun for now it must somehow escape from the mound. There is no help from its parents. The chick lies on its back and scratches and scratches at the debris above it. It is a difficult and exhausting task. The 60-centimetre journey out of the mound takes about two and a half days.

The rainforest's other megapode, the scrubfowl, is a warm olive-brown and grey in colour. It has only a short tail and instead of a show-off wattle it has a modest crest. There is no antagonism between the sexes, and when you see scrubfowl it is usually a pair foraging amicably side by side. Both sexes tend the mound. Unfortunately scrubfowl are shy birds and we do not see them very often. But they are very noisy, especially at night, when perversely they lose much of their shyness. They will come close to the bedroom windows to cackle and crow, in duet, in their loud gargling voices. They can keep you awake for hours.

Revolutionary

BULURRU, 2 APRIL

Geoff Tracey is a revolutionary in the sense that he, with a small number of others, revolutionised our perceptions of and thinking about Australia's tropical rainforest. It was he, with his senior partner Len Webb, who through extensive field work and study gradually came to the realisation that these forests were not recent derivations from Asia. The forests were, they discovered, quintessentially Australian and had been here ever since broadleaved rainforest existed.

The flowers of most rainforest plants are small and nondescript. Those of the primitive China pine, for example, look more like fruits than flowers.

In talking with Geoff, however, it is soon apparent that the social–political revolutionary is never far beneath the surface. But in his case the connection between the two kinds of revolutionaries is inescapable. What he and Len discovered was so important and the forests were under such pressure that a minor revolution was needed to rescue them from imminent and permanent oblivion. Geoff and Len were two of just a handful of people who knew, really knew, what was at stake here; how important this tropical rainforest was not just for Australians but for the whole planet. Both of them became revolutionaries out of necessity rather than doctrinaire political conviction.

Geoff arrives at Bulurru this morning, greyer and a little older looking at 61 years than when we last met 25 years ago, but unmistakeably the Geoff of old; unassuming yet eloquently passionate about the rainforest. He is dressed for the bush in working man's dark green shirt and trousers and heavy black boots. His frame is as spare as ever and his light blue eyes take in the forest appreciatively.

I ask Geoff to look at my two favourite species of large trees. They grow on top of the ridge across the creeks. One is the tree with the hollow in its straight satiny-textured trunk and the other has a trunk that appears to be made up of a series of braided stems. Geoff has a hard long look at both of them. He feels the smooth bark of the tree with the hollow. The leaves are too high up to be of much help. We hunt around for fallen leafy twigs but in so dense a forest it is difficult to say from which tree a particular leaf or branch may have come. In the end Geoff recognises the tree as *Syzygium endophloium*, the roly poly or mountain satinash. Geoff says it is the largest of its kind he has seen. The even larger braided-trunk tree also needs close scrutiny. We manage to bring down some leaves. 'This is very interesting,' Geoff says. 'It is a species of *Dissiliaria*. One species of this genus, called hauer or lance wood, grows in southeast Queensland and several others have been found at Iron Range on Cape York. Another species occurs on Mount Lewis, and I think yours is the same as that. It's endemic to this part of the world, one of a whole group of primitive species with restricted distribution. It's one of those interesting plants associated with the ancient metamorphic soils like this whitish stuff we're standing on. Your place being so wet and having these soils at the headwaters of a major river makes it an important refugial area, so it has lots of these interesting species.'

Flowers of a mistletoe.

Back at the house we talk about old times and tropical rainforest. This year has been a poor one for fruit throughout the tropical rainforest. Many fruit-eaters such as hypsies and cassowaries suffered real hardships. I ask Geoff if this is a regular, cyclical occurrence. 'We don't know,' he says. 'The only studies made on the fruiting and flowering of rainforest trees have been short term; only a few years. We know that certain species fruit just about every year and others only once every several years. We also know that there are "mast" years, as they are called, when certain species, especially the conifers, such as the kauris, produce extra-heavy fruit crops. But we have no accurate long-term records. The only observations ever quoted have been dug out of herbarium collections. They're the only ones we have, and what they tell you is where the botanists went and whether or not a thing was fruiting at the time they were there. These records don't tell you the reality. They create the illusion that we've got real data, because we've got them out of herbarium records. There is a certain amount of data there but it's notorious for being clumped around roads and easy access points. So the data are skewed.

'Even pretty thorough studies have their problems. For instance, we've done the work for the distribution patterns for the family Proteaceae, the silky oaks and things, and have come up with the result that says categorically, this is the pattern of distribution of the forty-five or so species of Proteaceae in north Queensland's wet tropics. The only problem with that is that the systematic sampling demanded by science means that you look at a number of pre-determined places. In the process you walk past a lot of stuff before you actually take your samples. So we proved beyond a shadow of a doubt, statistically, that this was the real world and that the pattern was like this. That's been published as a scientific paper. But in reality we walked past many species which we proved categorically weren't there. So science has a problem with its sampling system.'

Warming to the subject, an area of concern for many of us, Geoff continues:

'The purpose of all these studies is to understand nature, in this case to understand rainforest. It's the quest for knowledge and people's basic inquisitiveness in wanting to know. The concern I have is that so many people get hung up about numbers. My real interest is in interactions and interrelationships. It's much more revealing and also more fun. The statistical approach, that is probabilities and mathematics, is really about numbers. And as a result, by not using common sense in terms of what we want to know and how to go about doing that, we're throwing out the baby with the bath water. What's really lacking in the study of Australian rainforest are the classical, old fashioned natural history observations like those done by Charles Darwin, Alfred Wallace and lots of people since. We just have so little real data. And when we do get it we don't know how to keep it, because it's in people's heads and it dies with them. The ability of Bernie Hyland, Sam Dansie or myself to identify trees, for example, that will die with us. We don't know how to interpret it and how to keep it for other people. To teach someone you need to go into the forest over a long period of time and show them. Despite our technologies, it's still the only way. And we're not doing it.'

Despite Geoff's abhorrence of numbers, I do want to ask him some questions in that field. Using numbers may be just a gee-whizz way of looking at the forest, but it can help in gaining some understanding about its complexity, great age and the sheer volume of life; it can help in making it a little more comprehensible.

The flowers of the pink silky oak or tree waratah are exceptional in that they are large and colourful.

So I ask Geoff if he has any idea of the age of some of the larger trees.

'I have no idea how old some of the really large kauris or brush mahoganies are, they could certainly be more than a thousand years old,' he replies. 'As for the suggestion that some could be up to three and a half thousand years old, we can't tell which ones these are for nobody has really worked out any accurate ages. The trees that would be worth looking at for that sort of age are the *Leptospermum wooroonooran* [mountain tea-tree] on top of Mount Bellenden Ker. You know how long *Leptospermums* take to get to any size, well, there are ones up there with trunks a metre through. That is interesting but so is their relationship with similar species in Borneo and New Caledonia. They're old trees with old patterns of distribution.

'I think the great age of the trees is the sort of thing that fires people's imagination. Everybody wants numbers. My attitude is, if that's important to people then the World Heritage Management Agency has a responsibility to find out where the oldest tree in the wet tropics is and how old it is. There has been absolutely petty cash spent on questions of that type. It wouldn't pay for one timber truck, the amount of research money that's been expended on this whole area of finding out how old the forests are.'

Pushing my luck, I ask one more numbers question. I have been told, I say, that there are about 1200 species of trees in these tropical rainforests. Geoff's comment is:

'There are many more than that, but we don't know how many more. What we do know is that we're ignorant about the taxonomy of the tropical forests. I can only repeat that people are hung up about numbers. If they want to know how complex something is, they use numbers. The fallacy in that is that in China, for example, where they've done a lot of work on taxonomy [the science of classifying plants and animals], they may have many more species. That is because they've been working on it for a long, long time. Here when Bernie Hyland at the CSIRO did a piece of work on the rainforest laurels he increased the number of species by forty. He increased the number of *Syzygiums* [satinashes] by twenty-one. And that is just the work in only two groups. What I'm saying is that we do not know how many species of trees there are and if we counted them now we'd be wrong. What we do know, however, is that of the eleven or twelve hundred, or whatever number of trees there are thought to be, we in Australia have the most complex assortment of families and genera. Each genus has fewer species in it than the same genus in another country, but there is a greater diversity and complexity of groups, and many have more primitiveness about them. I think that's an important point. It's not just numbers.'

I ask Geoff if there are any rainforest ecologists or botanists who can go into the forest and name every tree species on the spot.

'There are very few people like that. Bernie Hyland would be the best. Tony Irvine, also of the CSIRO, and myself know the trees and other plants pretty well. I don't think any of us would know *all* the trees on the spot. For some species we'd have to take back some leaves or flowers, preferably both, and refer back to the herbarium collection. I've gone very rusty because I've been in politics [to do with World Heritage listing for the rainforest] for the last three years and have hardly seen the forest. You've got to be involved out there in the field all the time. If you let it drift and concentrate on other things then you've got to re-learn it. The interesting thing is that once you know how to identify the plants you re-learn extremely rapidly.'

I ask how he learned these things.

'It's a Gestalt thing, a way of looking at the world. It's the same as learning to recognise blue, black or yellow pencils, or pots or anything else. You learn to combine a whole lot of things instantaneously in your head. When you look at a tree you take in its size, if it has buttresses, whether the bark is rough or smooth, has it opposite or alternate, compound or simple leaves, and a whole lot of other things. Then you sort those out in your head and come up with a name. Just how you do that is different for different people. You find something, or a combination of things, that you recognise about a certain tree and you put a name on that. You're all the time building your knowledge of the trees' names and then you have to link the recognition in the field with the name.

'We always use scientific names. There's no other way to get to know the trees. The common, English, names are misleading and for most of the stuff there aren't any common names. Children have no hang ups about learning scientific names. They don't care what you call a tree, it's just a name and they learn these names as if they were learning a language. We put an adult perception on all this when we say it's too hard to remember *Cardwellia sublimis*, it's better to say northern silky oak.

'In the early days we used to come up with new things all the time. When I first started in this line of work, back in December 1949, there were eleven known species in the family Annonaceae, now we know there are at least thirty and there are probably still more out there. In those early days we used to find new species in most places we went. Those discoveries sometimes had wide-reaching effects. I well remember one day when we stopped for lunch north of the Daintree River at Noah Creek. Just while having lunch I found six new species of plants. Among these was the one I was searching for, *Idiospermum australiense* [ribbonwood or idiot fruit]. The fruit of this tree is the size of a cricket ball and fragments found in the stomach of a cow that died from eating it were sent to the Queensland Herbarium for

At certain times of year the pods of the scarlet bean, which grow low on the trees' trunks, light up the forest gloom.

identification. Dr Stan Blake recognised the fragments as *Calycanthus*, the first record since about 1875. Stan asked Len and I to look for the tree on our trip north. The ones we found that day had both flowers and fruit and turned out to be not *Calycanthus* but a distinct genus which Stan Blake named *Idiospermum*, the only survivor of a primitive plant family Idiospermaceae and unique to Queensland's wet tropics. Some of the other plants, such as *Gardenia actinocarpa* and *Noahdendron nicholasii*, are still only known from that one little area.

'At that time settlement and clearing had come right up to the south bank of Noah Creek and was about to swallow up the rest of the low lying areas including this refuge for primitive rainforest. It was only because Len and I recognised the importance of this place, in the early sixties, that we were able to stop the clearing and get a little national park established on Noah Creek. Out of that early action the whole idea of the immense importance of the greater Daintree region grew. Daintree became the focus of the conservation movement which led to the greenie bashing by the local politicians, the blockade of the Cape Tribulation road building and all that.'

I ask Geoff how he and Len Webb came to the realisation that these rainforests were Australian forests of ancient origins.

'It was fascinating,' Geoff exclaims. 'When Len and I started this we didn't know how to think or talk about rainforests. There was such a thing as just plain rainforest and then there were vine scrubs and dry scrubs, but that was all. We had to invent a whole system of classification and learn how to recognise similarities as well as differences between various kinds of rainforest. We had to go through the whole process of scientific discovery and classification in terms of all the forests we were dealing with, tropical and subtropical, lowland and upland, high rainfall and low rainfall and so on.

'My role was the taxonomy; to find out what families of plants grew together, which species grow exactly where; and how come that all the *Flindersia ifflaiana* grew around Kuranda and none around Boonjie where another kind of maple, *Flindersia brayleyana* is dominant. We also found that there was one group of *Flindersia* in north Queensland and another in the south. And as we went west into the dry country we ended up with a single species of *Flindersia* which grows as an isolated tree in an area near Broken Hill which has just 225 millimetres of rain a year. That kind of pattern really started to make us think about, and try to unravel, this whole business about the origin of Australian rainforest. When we looked at it from that perspective, the British botanist Hooker's 1868 theory about it all coming down from southeast Asia just became utter nonsense.

'So we sneaked up on it like that over a long period of time. When we started out, though, it was not to understand the forest, but to exploit it for drugs. We were looking for new drugs in plants, that's how we got our jobs with CSIRO. But soon after we started the Australian government, in its wisdom, decided to let the CSIRO become involved in this new science of ecology. They let Len Webb have his head to study the ecology of tropical rainforest. Milton Moore studied the woodlands and Alec Costin the Snowy Mountains. They had a biome each and they were the first ecologists in CSIRO. From that the whole understanding of the Australian vegetation developed.

'It was tremendously exciting. Len and I had a long history at CSIRO [both are now retired], but we were always a tiny part of that giant research organisation. Our section never grew bigger than just the two of us. We spent our lives trying to understand the forests. Len has gone in a different direction now. He is a Professor in the Environmental Sciences Department at Griffith University in Brisbane. He's gone into a lot more of the philosophical aspects, things like values and a true understanding of nature as a whole. Whereas I, being a more pragmatic sort of guy who has a different background, I've been into the repair process. We must learn how to rebuild these complex systems for, as I see it, that's going to be one of the most important things in the future; actually to be able not just to plant trees and rebuild some kind of system but to understand the very succession processes that are going on. Man can then enhance the regeneration process in tropical countries. One of the major problems in the world is to rebuild tropical forests, or at least create the conditions where they can rebuild themselves. At the moment we don't know how to do that.'

I ask Geoff what his background is and where his connection with the bush had its beginnings.

'I grew up in Cairns. My first interaction with the bush was at Laura Station [now Lakefield National Park] which was owned by one of my uncles. That was during World War II when I was about twelve. I started to learn about the bush there, but that was cattle country not rainforest. My teacher was an old Aborigine called Bob Ross. He had all the tribal marks on him and a hole through his nose. He took me fishing for barramundi. He knew all the plants and told me what they were used for. That was, quite clearly, when I started to interact with the bush.

'I didn't go into any rainforest till after I'd left college and got this job with Len. My first real experience of the rainforest was when we were actually confronted with it, a green maze it seemed to me then, and had to find particular species of plants in this overwhelming place of plants.'

It seems incredible that someone who has done so much work in the rainforest and has had so much influence in its understanding and preservation, did not seriously set foot in it till he was 20 years old, even though he had lived right beside it. Geoff acknowledges that perhaps this is unusual. He adds:

'But you've got to look at my background, an Irish-Catholic background. First of all I went to the Brothers School. Then we had World War II when we were exported out of town for a while. After that I ended up in boarding school. My whole education was biased against anything to do with the natural environment. When it was all said and done I was an ordinary, everyday bloke, a product of the Australian culture of the post World War II era.

'Things changed when one of my cousins, who had been a flyer during the War, became a teacher at Gatton Agricultural College in southern Queensland. He was instrumental in getting me down there. He also taught me some botany. This job with Len Webb at CSIRO was one I just picked up out of the newspaper. So it was purely fortuitous that I ended up where I did and that Len and I, from completely different backgrounds, were able to work together.'

'What was Len's background?' I ask.

'His father was a station hand and horse breaker and his mother was a station cook, out near Barcaldine I think it was. Len was very bright. He topped his class in English and other subjects. In the end he went down to Brisbane with his mother and finished his schooling there.

'His first job was with what was then the Queensland Department of Agriculture and Stock. Just by chance he was assigned as clerk–typist at the Queensland Herbarium. He'd never even heard of a herbarium before that. C. T. White, the old government botanist, took an interest in Len. Not long afterwards Len began part time studies at the Queensland University and during the war joined the University Regiment. One day they said at a Regiment parade, "Anyone who has done first and second year chemistry step one pace forward." Next they said, "You men are going to make munitions." So Len ended up making munitions. From there he picked up this job late in the war, in 1944, as a botanist on a drug plant survey. Australia, apparently, had run out of drugs and the government was looking at plants as a source of new drugs. And then, of course, he went to the CSIRO. It was just one of these things that happen.'

So the son of a horse breaker and station cook from the sheep country of western Queensland and a country kid from Cairns turned scientific thinking about Australian tropical rainforest upside down. Through their work and their discoveries, backed by the force of their personalities, they not only laid the foundations that led to these forests being protected through World Heritage Listing but were themselves a major force in achieving it. It was the culmination of their lives' work. Without the catalyst their insights provided we probably would still think of our tropical rainforests as displaced pieces of Asia. We would be unaware of the fact that these forests are where most of our vegetation originated, that they are places where rainforests have grown as long as there has been rainforest on this planet. It is also not inconceivable that without their efforts in the conservation movement, which continue to this day, many species and whole forest types would have disappeared without ever revealing their secrets. The names Webb and Tracey will always by synonymous with Australia's tropical rainforest.

Green Possums and a Python

BULURRU, 6TH APRIL

A green possum eating the leaf of a rusty fig.

Much exciting animal activity takes place after dark. I go out regularly to investigate what the possums, owls and nocturnal reptiles and insects are doing. Last night as I left the house my torch beam surprised a small dingo. After a quick sidelong glance at me it trotted off, causing the terrified pademelons to thump their feet in alarm.

Only a few metres further on I picked up the reflections of a pair of eyes like burning coals in the darkness. I walked closer and discovered they belonged to a green possum sitting low in the rusty fig; the same tree that attracted hundreds of fruit-eating birds a few months ago. There was no fruit last night. The possum was eating the large rubbery leaves, obviously not minding the sticky white sap, perhaps even relishing it. I moved right under the possum until he was not more than two or three metres above me. Neither I nor the light disturbed him. He looked placidly down over a

leaf clutched in his right hand. With an air of calm detachment he soon continued eating.

The green possum was as perfect an individual as I had ever seen. His dense, soft fur was of an even length which gave him a rounded, 'cuddly' appearance. This roundness was emphasised by his short ears which barely projected beyond his fur. His greenish colour was created by the mixture of yellow, black, grey and white hairs in his coat. No individual hair is actually green. The possum was so close that I could clearly see his bare pink nose flanked by long shiny black whiskers. His large brown eyes had vertical pupils which is rare among Australian mammals. Touches of white around his eyes compounded his look of wide-eyed innocence. The possum did not have a hair out of place, not a toe-nail was chipped, there were no ragged edges to his ears, his eyes were clear and his moist nose without scars. In the rough and tumble of a rainforest where life can be harsh, the green possum always looks vulnerable to me. Its seeming innocence and slow-moving trustingness are incongruous in a place where large owls and snakes hunt and where more aggressive possums compete for space. Yet it must cope quite well, for green possums are by no means rare.

Later, on my return, I glanced up to see if the green possum was still there. To my surprise two of the possums looked down at me. One was still eating, but the other was curled up beside him in a typical resting posture. With one large hind foot anchoring it to the branch with a strong grip, it rested on its rump with its tail curled forward. When sleeping the other three feet, the tail and the head are all tucked into its white underside and the possum becomes a green ball of fur in a green forest. This species needs camouflage because unlike the other possums it sleeps in the open during the day and not in a tree hollow or nest.

Curled up asleep.

This morning the rain has cleared. The sun is warm. I can see the green possums' rusty fig from my kitchen window. I catch a slight movement among the ferns at the base of the tree. Something glints in the sun. An amethystine python slides out of the undergrowth and loosely coils itself in a patch of sunlight. It lies directly beneath the branch where the green possums were last night. The snake is not huge, perhaps three metres long. Amethystine pythons can grow to more than twice that length. However, it is in immaculate condition. Its satiny skin shines with iridescence in the sun. The sheen is mostly of a violet, that is, amethystine, colour. I am delighted by this new visitor ... but also alarmed. It brings to mind the experience of a researcher studying tree-kangaroos.

An essential part of the study was to follow the animals' movements. A number of them were caught and fitted with radio collars, which emitted signals the researcher could home in on. One fateful day a signal led not to a tree-kangaroo but to a large amethystine python with a bulge in its stomach. If the much larger, very tough tree-kangaroos are victims of the python what hope is there for my green possums?

This afternoon the snake vanishes into the undergrowth. I do not see it again. Night after night I search for the possums, and usually I see at least one in or near the fig.

For a few days I have noticed a marked seasonal change. Rain is still frequent, but it no longer comes in heavy monsoon downpours. They are replaced by light showers borne on southeasterly winds. Sarsaparilla vines, topaz tamarinds, brush mahoganies, hard alders and various kinds of satinash have put out bright red and pink new foliage. Certain walnuts, bollywoods, pepper woods, aspen and the spectacular bumpy satinash are in flower. Even on the occasional bright sunny day it is warm rather than hot. I see fewer reptiles. Scores of rainbow bee-eaters call in trilling voices as they move northwards on migration. The black-faced flycatchers, except for the occasional straggler, have also moved on. The channel-billed cuckoo and the koel will soon be gone. These birds will spend part of our drier season in Papua New Guinea and places beyond.

Night of the Possums

BULURRU, 10 APRIL

With darkness, a fine mist seeps into the forest. A few crickets chirp. Boobook owls call in the distance. Mist particles bind together into drops that patter onto the foliage and ground litter. The creek mumbles in the valley below. Otherwise it is silent on this windless night.

Armed with a strong spotlight, I am in the tall forest the other side of the confluence, looking for night animals. Darkness set in two hours ago. There was no point in being here much earlier, as some of the species of possum I hope to see are not early risers. Whilst the sky still held some light, a shrill and insistent chorus of cicadas started up as they do every evening at this time of year. The insects are sometimes called possum alarms. But many a possum sleeps right through it.

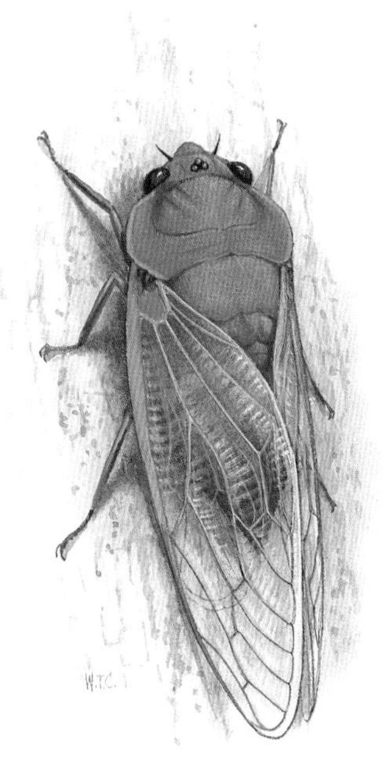

The cicada Cyclochila virens *sings loudly at dusk, a time when most possums come out of their dens. It is, therefore, known as the possum alarm.*

For four hours I walk along Bulurru's trails probing the forest with my light beam. Most of the animals will look at a light; when they do, their retinas reflect it, making their eyes shine with surprising brilliance. Usually this eyeshine is red, but the eyes of one possum reflect a yellow or green light, depending on the angle of the torch. I find most animals by picking up the glow of their eyes, sometimes from as far away as a hundred metres. But scuffling noises on the ground, scratchings on tree bark, the sound of teeth gnawing on hard seeds and shaking leaves also lead to interesting discoveries.

I set off down the main trail along the ridge. Insects dance in the light

The amount of white on a Herbert River ringtail's arms and chest is different for each individual.

beam and are caught on the wing by bats. Time and again I hear red-legged pademelons bound deeper into the forest, but I cannot see any. Low in a tree up ahead I pick up red eyeshine; always a moment of excitement. What animal is it? What is it doing? I need to be a lot closer to be able to tell. Keeping the light steadily on the animal, rather unkindly dazzling it, I slowly move closer as silently as I can. When about ten metres away I stop. Blinking at the light is what at first looks like a miniature panda: a black animal with a white chest, flanks and underside. Both arms are also white and have a black triangle at each elbow. The thin tapering tail, held coiled like a watch spring, is black with a white tip. I creep still closer. The Herbert River ringtail possum keeps looking at the light. His nose is not pink but purplish, his whiskers are just as long as those of the green possum, and his eyes are brown. No two of these possums have exactly the same pattern, but I have never seen one before with so much white on it.

What look like short pieces of white thread stick to his nose and fur. So hard had I been concentrating on the possum that I had not taken in his surroundings. I realise that he is sitting on the knobbly protrusion of a tree trunk which is covered in creamy white flowers. The possum has been eating the flowers of a bumpy satinash and their stamens are all over his face. He still has part of a flower firmly gripped in one hand. The tree's trunk from ground level to ten metres up is covered in bumps and each bump sprouts a profusion of flowers. Slowly I back away so as not to disturb the possum unduly and to let him get on with his meal.

The exact sequence of events from now on is a little blurred; so much is happening. I gather a series of vivid impressions, cameos of animals caught in a pool of light in an otherwise dark forest.

I see so many more Herbert River ringtails that I cannot keep count. One is almost entirely black with just a small white star on her chest. Another has white bands around her arms, a third has one white hand and wrist while the others are black. I surprise one while he is eating the leaf of a candlenut tree. He sits frozen in the act of putting the leaf in his mouth, a leaf so large it almost hides his face. After a few seconds he resumes eating, taking large bites out of it while holding it in both hands.

The fourth or fifth Herbert River ringtail sits low in a vine, staring intently in my direction. I walk slowly up to it, putting each foot down as carefully as I can. Inevitably some leaves crunch and a twig snaps loudly. Not in the least panicking, the possum climbs up the vine in a fluid motion that makes her appear to flow along. Without a pause she grabs the outer foliage of a sapling, transfers her weight while briefly anchored by a coil of her tail, and is gone. It all seems very casual and slow until you realise how quickly she disappears from view and reappears high in a distant tree. Only rarely will these ringtails leap and when they do it is for only a short distance.

Green possums, of which I see only two and not very closely, move with the same easy grace and deceptive speed.

Right beside the trail I hear a faint rustle and the sound of a leaf being torn. By moving around a dense mountain mangosteen bush I see a dark

Leaf-tailed gecko.

The striped possum rips open decaying wood with its long sharp lower teeth and then extracts the wood-boring insects.

gular head, just emerging into an open space between two trees. At my approach the snake raises its head more than a metre off the ground and flickers its tongue. Slowly it lowers itself again and continues its stately progress. The python is close to five metres long and takes quite a while to clear the path in near-silent stealth. It is so much larger than the python near the fig tree, and out here in the dark forest looks far more sinister.

There is nothing subtle about the ripping and tearing sounds that come from a little way into the forest. All I can see of the striped possum when I shine my light in its direction is its bushy tail, half black, half white, sticking out of a shallow cavity about three metres up. The possum sits there motionless, waiting for the light and the noise to go away. But I make myself comfortable between two tree buttresses and switch the light off. The mist has become a fine drizzle that settles in beads on my hair and clothes. After what seems hours, but is probably no more than 15 minutes, I hear the possum stir again. I wait till it is ripping wholeheartedly into the wood then switch the light on again, but point it in such a way that only the faint light of the beam's periphery illuminates the marsupial. It stops, backs out of the cavity and looks back over its shoulder. Apparently reassured, he resumes his excavations. First he sniffs at the decaying wood and drums on it with his long sensitive fingers, probably locating some borer or its larva. Excitedly he tears into the dead tree with his teeth, making the chips fly. Every so often he pulls out some insect and, crunching noisily, eats it. Sometimes he pauses to probe a borer's tunnel with quick urgent movements of his elongated fourth finger, either to pull out the insect or perhaps to locate it accurately. The possum's actions are so fast that it is difficult to see exactly what is happening.

I find my way back to the trail. As I shine my light along it, a bird flies up about 20 metres on. When I reach the spot, the sooty owl has not flown far. It sits looking at the light from a branch only about four metres high. In its talons is a smallish white-tailed rat, too heavy for it to carry any distance. This larger relative of the ubiquitous barn owl is finely marked in dark grey and white. The bird does not fly off, away from the intrusive light, it merely turns its back and begins to dismember its prey.

Back on the slope above the confluence, high in a tall tree, I hear the familiar coughing and hissing of quarrelling brushtails. These rough, tough, rumbustious possums are the very antithesis of the delicate and graceful ringtails. The muscular heavyweights fight and leap, scamper up and down trees, eat almost anything and are into everything. The spotlight's second battery for the evening is weakening. I can only just make out the possums' vivid coppery-red and shining fur through the mist drifting across the tree tops.

It has been an exceptionally eventful evening. I sit down on a moss covered log and shine the light around one more time. Not five metres away a diadem bat, large for an insect-eating species, hangs from the end of a dead twig. It twitches its pointed ears and extraordinarily convoluted nose in my direction.

I switch the light off. Now I am in the animals' realm. It is very dark.

There is no moon. The clouds and mist obscure most of the light from the stars. For a wide array of animals, possums, bats, frogs, lizards, owls, snakes, crickets, moths and many others, night time is the time they function best —by scent, by hearing, by sonar, by touch and by extraordinary eyesight.

Zodiacs and Tubenoses

Bulurru, 2 May

Most of the hazelwood trees are covered in bunches of flowers. Their scent is heavy on the air. A few trees have already finished and the dark brown carpet of dead leaves beneath them is covered with white petals which sparkle after a passing shower. The hazelwoods near the back verandah are at their peak. From the top of their ten metre high crowns to the lowest branches sweeping almost to the ground, they are covered in cascades of dazzling flowers. The day-flying zodiac moths swarm over them in their hundreds. They flutter around the clusters then settle for a few moments to drink nectar, creating an ever-moving shimmer of the bronze and dark blue on their upper wings and bands of pale turquoise on their undersides.

During a period of sunshine between showers I try to take some photographs. While adjusting the camera, I notice a procession of shadows on the ground; more moths arriving. But one shadow is very much larger than the others. I look up quickly. A male birdwing butterfly, huge among the zodiacs, lands briefly on a flower. He must have newly emerged from his pupa for he is perfect in every detail, his bright emerald green, yellow and black wings do not seem to have a scale out of place. He is immediately surrounded by a score of fluttering zodiacs. Soon he flies off again. Few other insects come to the nectar. A pair of shining blue Ulysses butterflies, looking in their fast zigzagging flight like flakes of sky falling down, pay a brief visit, but they also quickly zoom off again.

The zodiac moths, besides flying by day, resemble butterflies in other ways. In size and shape they look very much like swallowtails, complete with tail-like projections on their hindwings. But other anatomical features, such as antennae that end in fine points instead of club-like knobs, put them clearly with the moths. The distinction between the two kinds of insects sometimes becomes blurred.

In the afternoon black clouds roll in again from the southeast, making a wonderful contrast to the white flowers. Soon heavy rain comes down. The zodiacs continue to feed on the flowers. The rain runs down their wings and drips from their 'swallowtails'.

By late evening the rain clears. The air is still and fresh. Leaves sparkle and drip in the late sun. A friend and I walk towards the waterfall which has

Common or scaly tree ferns.

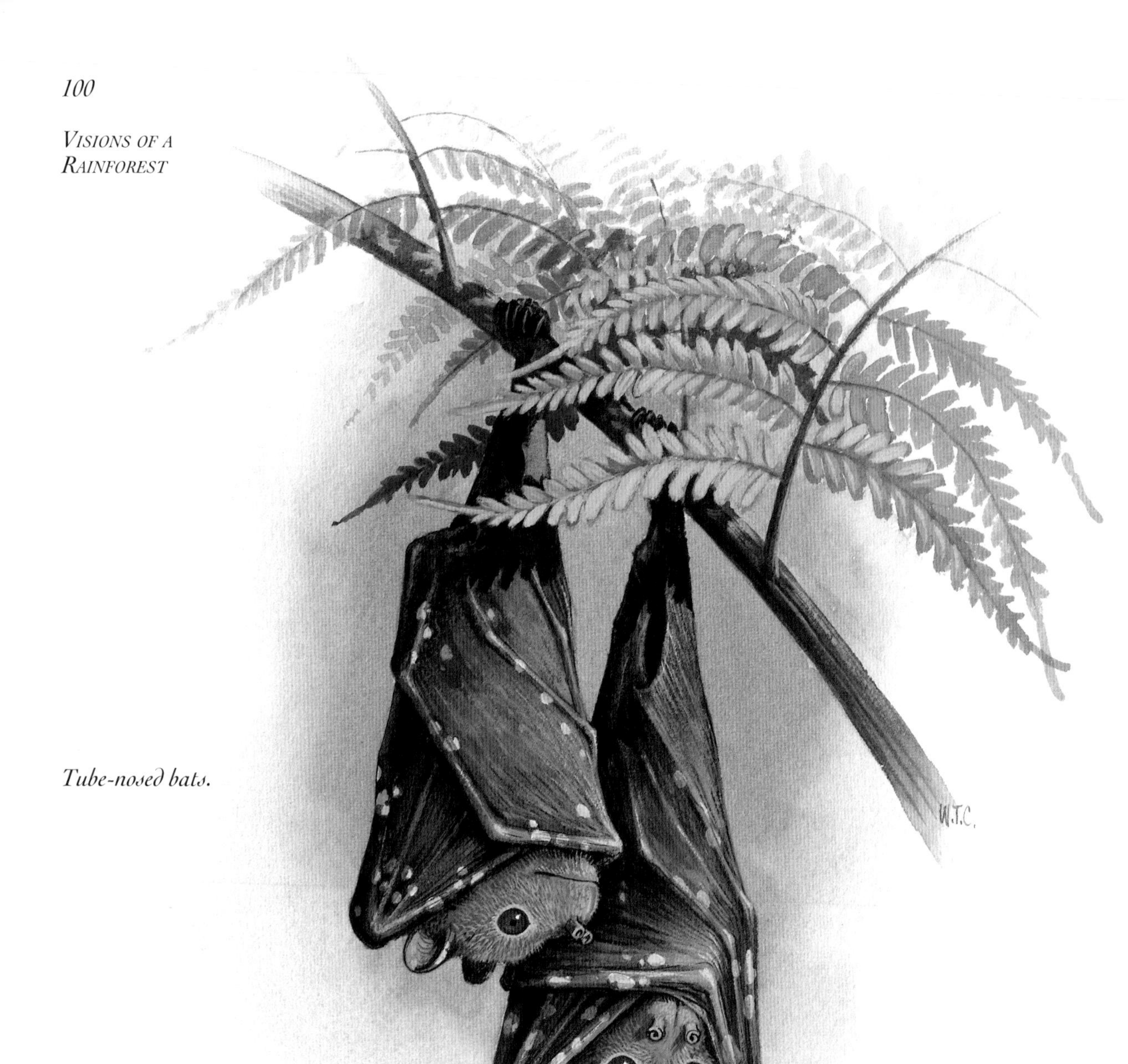

Tube-nosed bats.

become much noisier after the recent rain. We stop at the fern gully and admire all the different kinds of ferns, from ten metre tall tree ferns and a climbing species that reaches twice that height in surrounding trees to tiny ferns growing as epiphytes on a laceflower tree. My friend spots something interesting in the nearest tree fern. It looks like a bunch of dry leaves to me. She asks if what she sees could be flying-foxes, so I grab my binoculars for a closer look.

The two bats hanging there cuddled closely together are indeed flying-foxes, of a sort. They are much smaller, however, than the mostly black species commonly seen in flowering and fruiting trees, and are pale brown in colour. On their wings and ears they have yellow spots, giving them the appearance of dead leaves dappled by sunlight. This pattern affords them almost perfect camouflage. Each clutches the frond of the tree fern with one foot; the other is tucked under its wings which are lightly wrapped around its body. They look at us through slitted brown eyes. Water droplets adhere to their soft, bare, leathery wings. Each nostril is elongated into a tube several millimetres long. We are looking at tube-nosed bats. Why the snorkel, is the inevitable question? these bats, while not rare, are seldom seen and have been little studied. It was thought they were fruit-eaters and that the tubes allowed the animals to breathe as they buried their faces in a ripe pulpy fruit such as a Davidson's plum. The observations that have been made of captive tubenoses do not bear this out. Their preferred diet, it was discovered, is not fruit but nectar and pollen. The need for a snorkel remains a mystery, like so much in these forests.

The pair remain at their daytime roost in the fern gully for fifteen days.

King Parrots and a 'Butterfly' Happening

Bulurru, 15 May

Early morning is crisp. The skies are clear and Mount Bartle Frere stands sharply outlined. There is a wintry quality in the air, a clear signal that the cyclone season is over. We have been fortunate for none has crossed the tropical coast. Several made threatening approaches but veered south at the last moment.

The dew is heavy on the grass and drips from the trees. My boots are saturated by the time I have walked to the narrow ridge that overlooks the World Heritage rainforest. When the sun finally clears the mountain it is instantly warmer.

Not all moths fly only at night. These three species are out during the day. In the centre is the zodiac moth which resembles a swallowtail butterfly. On the left is Dysphania fenestrata *of the lowlands and on the right* Milionia queenslandica *which brightens Bulurru's forests.*

Tree-Climbers and Stingers

MOUNT BALDY, 16 JULY

Sunrise is a good time to look for tree-kangaroos, or tree-climbers as they are known locally. I am in Mount Baldy State Forest, right on the crest of the Great Dividing Range and on the edge of the rainforest. The track follows the line where it meets the drier eucalypt woodland. In the east the sky gradually lightens and the mist that had been hovering over the trees soon dissipates. It is chilly. I find the tree-kangaroo by its silhouette, framed by leaves, against the warm glow of the sky. It is still dark enough for my spotlight to be needed to see the animal's somewhat stolid, round black face peering down. Its bare nose too is black and leathery like that of a koala. The ears are small and rounded. With the strong claws of one hand it holds on to an upright branch. The other fist is clasped around the basswood leaves it was eating. Its long tail, which is not prehensile, hangs down like a vine. This Lumholtz's tree-kangaroo, to give it its full name, is a large male. He stands on a horizontal branch and I can clearly see the kangaroo-like pads on his hind feet. After staring down for some minutes he shuffles along the branch towards the tree's trunk. I walk closer, for a better look. When I am almost under his tree, he suddenly leaps down, about ten metres, hits the ground with a loud thud and bounds off into the forest. This arboreal kangaroo still feels more secure on the ground. It also lacks the fluid grace and tree-climbing finesse of the ringtail possums, but it more than makes up for this in sheer strength and toughness. Years ago I saw tree-kangaroos leap from even greater heights, 15 to 20 metres, without coming to apparent harm.

As the day brightens a few birds call. I can now see the sharp demarcation between the two types of forest. On one side is the thick, dark, green rainforest. On the opposite side of the track are the eucalypts. The trees are equally tall but their crowns are sparser and wider, the foliage a paler grey-green. Some trunks are covered in shining white bark, others are dark and fibrous. Few epiphytes have gained a foothold. The understorey is mostly wattles and small-leaved shrubs rising out of grass. Here and there a small rainforest tree has managed to establish itself. The rainforest might invade these more open woodlands except that most years they are burnt. The fires, which are no accidents, destroy the thin-barked rainforest species but leave the eucalypts unaffected. If left undisturbed, rainforest would grade into the eucalypts instead of facing it along a sharply defined boundary. Their area of transition would be narrow, no more than half a kilometre, for here at the summit of the Great Divide rainfall falls off abruptly along the western slopes.

The sun rises over the tree line. Up ahead leaves rain down out of a tree, although there is little or no wind. It is the work of another tree-kangaroo and her well-grown young. But why are they pulling off the leaves? I

Lumholtz's tree-kangaroo.

Elkhorn fern. Tree-kangaroos often use large epiphytic ferns like these as their daytime retreats.

do not want to disturb them, so I stop while still more than 100 metres away. Using binoculars I can see what is happening. Both mother and young pull leafy branches of the brown bollywood towards themselves then bite off, and eat, just the leaf stems, letting the leaves themselves drift to the ground.

I find a convenient log, in the sun, to sit on and watch the animals. After a few minutes the young tree-kangaroo, which is much fluffier and woollier than its mother, curls up asleep on a branch with its rump resting against the trunk. It tucks its head and front paws into its chest.

About ten metres away a shaft of sunlight highlights a tall shrub with heartshaped, hairy leaves and beautiful translucent pink-purple fruit. A catbird flies in and eats some of the fruit, hopping from one hairy leaf-stalk to another. It seems to suffer no discomfort from either eating the fruit or moving about the stems. That is a surprise for the shrub is the dreaded, even dangerous, Gympie stinger. It thrives along the edges of the rainforest and in places of major disturbance. I have been stung in the past and the pain is severe. There is no antidote. The sap of the cunjevoi, a large-leaved aroid, is recommended by some people. It does not help and is, in fact, a skin irritant itself. An Aboriginal friend said they used the sap of the wild banana but added that it did not really work either.

The hairs on the stinger are made of silicon so in effect are minute slivers of glass. Each is tipped with poison similar to that of snake venom. When you brush against a leaf or stem the hairs pierce your skin, break off and stay there. That is why the pain can persist for months. The Gympie stinger's poison is just one weapon in the constant war between plants and the animals that eat them, especially the insects. But not even the strongest poisons make the plants immune from the attack of all animals. Several large grasshoppers are gnawing the stinger's leaves and there is a butterfly, called the white nymph, whose caterpillars only eat the leaves of the various kinds of stinging tree.

The young tree-kangaroo has woken up and joins its mother. The sun is on them and they seem to be getting hot in their dense, silky fur. The female, unlike the male earlier this morning, does not leap to the ground but slides, tail first, slowly down the bollywood's trunk, using her hands as brakes and sending down a shower of bark. The youngster remains up the tree and continues to feed on the leaf-stalks as its mother hops off into the cool forest. I walk quietly towards the tree. The young, its round black face surprisingly bear-like, looks down for a moment, then it too descends.

Fruit and leaves of the Gympie stinger which can cause excruciating pain. The caterpillars of the white nymph butterfly eat the stinger's leaves, despite their poison.

Winter Interlude

BULURRU, 2 AUGUST

Things have been quiet over winter. Cloudless nights have been cool, on occasions even cold, the days often sunny and warm. This is the first of the drier seasons. But even now low clouds and mist sometimes roll in from the southeast and light rain falls. Nights are warmer then and coax pythons and brown tree snakes from their hiding places.

During the night it rained and this morning a light, warm drizzle persists. Small black ants open up the entrances to their nests, building miniature pyramids of red soil on the lawn beneath the rusty fig. I know from past experience that the ants are aggressive and when you accidentally

stronghold of more of the same kind of fig. The large spreading trees, which are stranglers, are variously known as the Australian banyan, deciduous fig, white fig and, in one particular form, as the curtain fig. Here most grow half on land, half in the water.

To counterbalance the trees' list out over the water a web of intertwining roots, like cables and guy ropes, anchor the structure to the soil. Among these roots, between the pillars and where the pillars and branches join, there are all kinds of pockets and hollows. Fallen leaves have accumulated there, decayed, formed humus, and so created places for all kinds of other plants to grow. An enormous, spreading tree is a veritable garden of epiphytes and vines. The fig itself has shed its leaves. But it will not be bare for long. All its twigs are covered with pinkish-green leaf buds about to burst. For the moment, however, its nakedness only emphasises its covering of ferns, orchids and all kinds of climbing and clinging plants.

An umbrella tree grows as an epiphyte on the trunk of another species.

In the recesses where the roots come together, bird's nest ferns have taken root and spread a profusion of deep green fronds. Small tongue-shaped ferns march up the trunks, marbling them with their roots. Climbing pandans have crawled from the shore and smother some of the lower branches. Where two stout branches bifurcate an umbrella tree has established itself. The largest, thickest branch, reaching for the light that streams in from the lake, carries the heaviest burden of epiphytes. Between the ferns, clumps of liparis orchids push out stems covered in yellow-green flowers. Another kind of orchid wraps the branch in a network of roots. Its leaves hang down in tiers of dark green cylinders giving it the name of pencil orchid. Many of these orchids, hanging down like wind chimes, are covered in whitish, sweet scented flowers.

While I am admiring this tree and its associates, a pair of Atherton scrubwrens sings and twitters around me. I watch them more closely as they come near and notice one of them picking up some moss and flying off with it. The pair must be building a nest not far away. Wriggling through the vines, which are rampant right along the lake's edge, I try to follow the tiny birds. Progress is too difficult and the birds soon elude me. I have gone only about twenty metres but getting back to the path is quite a struggle. It is a bit like fighting yourself out of a large, coarsely woven bag. Vines grab me by the ankles, wrap around my knees, hold me by the waist and push against my chest. The hooks of a wait-a-while latch on to my ear. It is the attractive fishtail wait-a-while, but still a nuisance.

The stems give a little, but the harder I push the firmer they hold me. The only way out is the slow way; disentangling myself from one hooked tendril, one noose, one coil at a time. I console myself with the knowledge that tangles of vines like these are very unusual and not a constant feature of rainforest as so often imagined.

Free of the vines, I am conscious of the whistling, twittering sounds of a large gathering of wandering whistling ducks. Hidden by the lakeside shrubs and trees, I am able to approach the birds closely. Several hundred are noisily whistling, splashing, bathing, preening and socialising. A long line of them is perched on a low, thin branch of yet another deciduous fig tree. Every time another duck joins them, it makes the branch sway and the whole line teeters back and forth to keep its balance.

I hate to disturb this lively gathering but the path leads out in the open, close to the ducks. I could not face another round with the vines. As slowly and unobtrusively as I can, I walk out of the forest. Splashing loudly, the peculiar whistling of their wingbeats joining that of their voices, the ducks take off. They land again only about 30 metres away. They watch me walk by in total silence. As soon as I am out of sight their whistling conversations, their bathing and splashing resume.

Huge trees flank the trail that now rises up a slope. Two large red cedars, their trunks with dark bark, scaly like a crocodile's back, buttressed and brooding, rise out of the wait-a-whiles. These would have been valued trees even sixty years ago, before Lake Barrine was a national park, yet they remain, as do even larger blush alders and tulip oaks. We probably have to thank a farsighted forester from the past for this.

Barrine's most magnificent trees are neither red cedar nor tulip oak, blush alder nor fig. I have nearly come to the end of my walk around the lake when I am suddenly confronted by two gigantic bull kauri pines. These are not flowering plants but conifers. These two trees, standing only two metres apart and of about equal size, are truly enormous and imposing. I am struck by their vigour and perfect symmetry. 'Vigour' is a relative term here and refers to their health rather than their rapid growth. Their growth is slow. During the last 30 years the trees' diameter, at breast height, has increased by only six centimetres. Other very large trees in these forests are usually damaged and scarred by storms and cyclones. Most have hollows in them. Not these kauris. The 50 metre tall trees are absolutely unblemished and have wide crowns. It is difficult to establish the age of such a tree. Five hundred years is not an unreasonable estimate. Some people speculate it might even be 1000.

The most remarkable thing about the kauris, however, is that they belong to the same genus of trees that grew over most of Australia 175 million years ago. This was millions of years before the appearance of the first flowering plants. At that time most of Australia was moist and warm. There were vast forests composed of kauris, brown pines and treelike cycads. All three of these occur together at Barrine, survivors from ancient times and today growing within modern forests of flowering plants.

The giant kauris' roots are no doubt nourished by the lake's water, only a few metres away. The lake is fed only by the run-off from the surrounding slopes. Virtually no soil nutrients are washed or leached into the lake. Tests have shown that its waters, and those of neighbouring Lake Eacham, are among the freshest known in Australia. They contain few impurities and consequently few nutrients for aquatic plants and animals.

W.T. Cooper - 92

Wandering whistling ducks at Lake Barrine.

Fish swim along the clean bottom, however; catfish, a fish with the unflattering name of fly-specked hardyhead, and several others. While no creek enters Barrine, it overflows into a small stream. This, more than likely, was the route by which the fish reached the lake. Three kilometres away is slightly smaller Lake Eacham which is a completely closed system. It has fewer native fishes. The mystery is that it has any at all. There were once three species: the same fly-specked hardyhead, the northern trout gudgeon and the Lake Eacham rainbow fish. The rainbow fish was unique to Eacham and it was plentiful. Then other fish were introduced into the lake—sooty grunters, archer fish, Queensland bony bream and the mouth almighty. By 1989 the Eacham rainbow fish had disappeared. It is now extinct in the wild, the mouth almighty being largely responsible for its demise. Fortunately this rainbow fish lives on in the aquariums of the Queensland Government Fisheries Centre on the Tableland. Efforts to reintroduce it into the lake have so far been unsuccessful.

When looking at these singular lakes the question that comes to mind is: How did they come about?

The Dyirbal Aboriginal people know Barrine and Eacham as Barany and Yidyam respectively. The people say the lakes were formed by the rainbow serpent. The wrathful mythical snake blew up hurricane winds and cracked and twisted the very earth. Billowing red clouds filled the sky. People trying to escape the vengeance of the rainbow serpent were swallowed up by cracks in the ground opening up before them. During these upheavals the lakes were formed. It was a time, the Dyirbal say, when the Tableland was covered in open eucalypt forest with a grass understorey. There was no rainforest then.

The white fella story has some remarkable parallels. Geologists say that about 95,000 years ago there was still a lot of volcanic activity close to the surface on the Tableland. At Barrine, Eacham and a few other places groundwater percolated down to the hot substrate. Superheated steam was trapped between spaces in the rocks. Pressure built up. Eventually this could not be contained and enormous explosions blew out the craters. Lava and ash were spewed out and built the rims around them. When volcanic activity died down, the craters filled with water and became lakes which are now 65 metres deep. Rainforest from the surrounding areas subsequently invaded the slopes and grew to the water's edge.

Deep closed lakes like these are never flushed out. Whatever falls in them sinks to the bottom and decays. Some of the organic particles are preserved. If these could be identified, they would give a clear picture of past life in and around the crater. As it happened, microscopic but distinctive and identifiable particles of plants have been preserved at the lake bottoms since earliest times. These are the pollen grains of flowers. Many pollens can be identified as to the family of plants they belong to. Some can be taken further, to the level of genus, while for a few it is possible to determine the species.

The analysis of pollen grains retrieved from the bottom of Lake Barrine

A red-legged pademelon, a small rainforest wallaby, cleans her face.

shows that it was not always surrounded by rainforest. About 25,000 years ago the Tableland became drier and rainforest was replaced by eucalypts and grasses. Only about 10,000 years ago, a period spanning not all that many generations of kauri pines, did rainforest return. From the arrival of the first pioneer species to a more or less stable, well-established rainforest took 1500 years.

But where did the rainforest survive during the dry period? It retreated to the wetter regions, to Mounts Lewis and Bellenden Ker, to the headwaters of the Mulgrave, Russell and Johnstone Rivers, to the Cape Tribulation area and a few other places. Over the aeons there have always been such retreats during periods of dryness or volcanic activity. But when conditions allowed, rainforest returned. Only man has the capacity to destroy them permanently.

Platypus

Bulurru, 27 September

From time to time the platypus climbs out of its pool and grooms itself.

A strong northerly breeze has brought an unusually hot and humid day. The setting sun is nearing the tree line when I set off for the pool above the waterfall. I notice that a large clump of cymbidium orchids is pushing out bunches of flower buds. It is twilight by the time I reach the sandy edge of the pool. The opposite bank is a sheer cliff, its rocks obscured by cascades of glossy dark green ferns, shrubs and tall trees which have wedged their roots in cracks and crevices. On my side a latticework of burny bean vines hangs from the low branches of a large-leafed tree. Here and there hang clusters of the vine's pale green flowers. These will develop into large pods covered in brown hairs; hairs that will stick into your skin and give a burning sensation.

Whipbirds, hiding in a nearby clump of wait-a-whiles, give loud explosive calls. From high up the opposite slope comes the distinctive at-the-bower whistle of a satin bowerbird. I would love to find his bower and watch his rituals and displays. But it is too late to go and investigate. The stream water burbles gently as it enters and leaves the almost circular pool. Giving its high pitched call, an azure kingfisher darts across in a streak of blue.

Suddenly I notice something brown speeding up from the dark depths of the pool. A platypus surfaces almost at my feet. It is a small individual that either does not notice me or is undeterred by my presence. Floating, it stays in one place with occasional lazy movements of its webbed front feet. Its dive must have yielded a good harvest for it chews its catch, which it stores in its cheek pouches while under water, for quite some time. 'Chewing' may not be the right term as the adult platypus has no true teeth. Those are shed while the animal is still very young and are not replaced. The shrimps, dragonfly larvae and other aquatic animal life it dredges up are ground and cut up by ridged plates at the back of its mouth.

Having disposed of its catch, the platypus shuts the valves on its nostrils, closes the grooves where its eyes and ears are, and dives steeply down, leaving a trail of bubbles. I can just follow its movements. Sweeping its highly sensitive beak back and forth through the dead and decaying leaves at the bottom of the pool, it swims among logs and probes the narrow spaces and crevices between them with urgent and energetic movements. So deeply does it squirm into the confined spaces that I worry about it getting jammed.

Clouds of mud are stirred up and I often lose sight of the platypus. No wonder it shuts its nostrils, eyes and ears when diving, they would only get clogged up with silt and are of no use in that murk. The skin on its beak is studded with receptors so sensitive that few of its prey actually go undetected. Touch is the only sense organ the platypus needs while under water.

Flowers of the burny bean vine.

After a few minutes it rises to the surface again, propelled vertically by its strong front feet. The hind feet, which are also webbed, are pressed against the tail and are not used in swimming. Time and again the little animal dives.

The platypus is one of the species that symbolises all that is different, special, wonderful and exciting about Australia's wildlife. It is also a shy and retiring species only rarely seen. But, like so many animals on Bulurru, the platypus here is trusting and confiding. I see platypus quite often, at close quarters and at any time of day. On one occasion I saw two together in a pool in the sunshine at midday.

The last glimmers of twilight hover in the undergrowth by the time I walk back. The sweet yet spicy scent of the flowers of the aptly named spice bush is heavy in the air. Flowers of the hairy gardenia drift like white stars in the semi-dark. Something else small and white lies on the track. I pick it up. I can just make out that the white is the inside of the empty shell of a riflebird's egg. A young must have hatched in a nest not far away. I will have to come back tomorrow to try to find the nest, and also the bower of the satin bowerbird.

Phantoms

Bulurru, 28 September

In daylight I can see just how beautiful the riflebird's eggshell is. The base colour of its lustrous surface is pinkish buff streaked with rusty brown, dark brown tinged with purple, and purplish grey.

I have not gone far on my return to the platypus pool when I notice a cassowary dropping right in the middle of the trail. It was not there last night, and it is still very early. I am sure it is still warm. The giant bird must be very near. I look carefully around but can see nothing. We are in the middle of a dry spell, a time when leaves crackle noisily underfoot. It is almost impossible to approach any animal unnoticed. When it is damp the leaves are soft and pliable and you can walk very quietly. I listen for the cassowary's foot-falls but cannot hear them.

The search for the riflebird's nest is equally unrewarding. The female, who builds the nest unaided by the male, usually places it in a thick tangle of vines, in an epiphytic fern or in a clump of upright shoots in a tree. The outside of the nest, which is made of twigs and vine tendrils, is decorated with small fern fronds and often with the sloughed skin of a snake. When the young hatches the female takes the empty shell and drops it not far away. So a nest must be somewhere close to where I found it. I examine every tangle of vines, every large fern clinging to a tree, every dense bunch of leaves. I find old nests of catbirds and shrike-thrushes and that is all. After about an hour I have a painful crick in my neck and abandon the search.

How different the pool looks on this sunny morning. All around it trees lean over the water. The sun filtering through the foliage flecks the dark

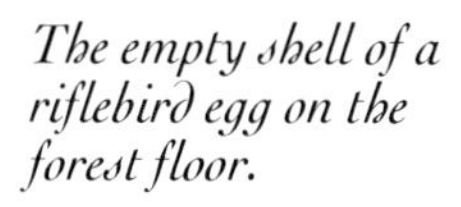

The empty shell of a riflebird egg on the forest floor.

water with patches of brightness and catches the pink, rusty red and pale yellow-green of the trees' new shoots. Old leaves are being shed and when a breeze springs up green, yellow and red spent leaves rain down. The current carries them away and deposits them in back eddies and still corners.

As I look at a particular raft of these floating leaves, a platypus suddenly pops up in their midst. A close look through binoculars soon establishes that this is a different individual to that of yesterday. It is much larger, with a broader bill and more robust tail.

Where a patch of sunshine illuminates the water something very slowly floats to the surface, much too slow to be another platypus. It is a sawshell turtle, a large one with a shell about the size of a dinner plate. Another tiny one floats up with it. When they reach the surface and gulp some air, the small one swims to the other's face and begins to nuzzle it. I cannot imagine what this could signify for turtles do not look after their young nor are they known to show much affection for one another. The larger turtle ignores the attention. Soon the two submerge into the pool's depth.

All morning I have heard the pleasing resonant song of the black butcherbird. It is Bulurru's most pervasive voice; its theme song you could say, yet its owner is also its most elusive bird. I rarely see it, and when I do it is only a glimpse of a black shape in the darker parts of the forest.

I cross the creek above the pool. The opposite bank is a sheer rock wall for the first 15 metres or so, but there are enough tree roots and vines to make the ascent. It is like climbing a ladder. Once on top of the cliff, there is a long gentle slope to the top of the ridge. I feel quite excited for I have never been to this particular place before.

The forest is particularly beautiful, with slender tall trees. Here and there vines have fallen in a heap or a large branch has crashed to the ground, but otherwise it is quite open. There is

A sawshell turtle comes up for air.

W.T.Cooper. -92

The buttresses of a large brush mahogany or red carabeen, in mature rainforest. On the left a Rhaphidophora *vine, a kind of aroid, climbs up a tulip kurrajong tree. Behind them, to the left, grows a cunjevoi, also an aroid. It was in forest like this that so many animals hid like phantoms.*

no need to struggle through the vegetation; I can concentrate on the animal life around me.

Ever since I got to the pool I have been listening for the at-the-bower whistle of the satin bowerbird. But all I have been hearing is the torrent of song of a tooth-billed bowerbird poured forth from high in a tree. The singer remains hidden. Like all tooth-bills he is a great mimic and mixed with his own powerful notes are imitations of those of figbirds, shrike-thrushes and flycatchers. This is the area from which I heard the satin bowerbird's calls last night. I search all around in an ever-widening circle. Nothing. It may well be that the satin's calls I heard were mimicked by this tooth-bill.

I find the nest of a chowchilla, a bulky structure made of sticks and dead leaves and lined with finer materials. It is as yet incomplete; the domed top is still missing. Or perhaps it has been ripped away by a predator.

Near the top of the ridge is a hollow tree. The interior is whitewashed with the droppings of a bird of prey. Half a dozen owl pellets lie on the ground. These are ovals about five centimetres long made up of tightly packed animal fur and bones: the parts of its prey that the owl cannot digest and which it regurgitates. I carefully pull some of the pellets apart. Several contain complete skulls of mosaic tailed rats, another that of a long-nosed bandicoot. Another is almost completely made up of bandicoot fur, and one contains the long black whiskers, skull and fur of a very young green possum. Surely these pellets were regurgitated by a rufous owl. They are much too large for the other two owls here, the sooty and the boobook. The hollow tree must be the owl's roost. I tap the tree, hoping to flush the bird out. I search the dense foliage surrounding the hollow, expecting to see a pair of yellow eyes staring down at me. No such luck.

Slowly I wander to the top of the rise. This kind of ridge top is the ideal place to find the bower of yet another bowerbird, the striking golden species. So far I have not found one on Bulurru, though I have been searching for three years. There are at least two bowers with attending males at Chowchilla and other neighbours also have these birds in their forests.

Once again I stop to have a good look around for the tell-tale pyramids of sticks piled against a small sapling. My heart skips a beat. I can just make out what appears to be such a structure. It is the work of a golden bowerbird all right, but it is only half a bower and not decorated. A complete bower has two pyramids of sticks, one taller than the other, joined by a horizontal perch. Such a structure is embellished with yellowish beard lichen on one side of the

An owl pellet. Owls regurgitate oval shaped packages of the fur, bones and other indigestible parts of their prey. This one contains the skull of a mosaic-tailed rat and the jaw, with fine sharp teeth, of a long-nosed bandicoot.

perch and the seed capsules of a small tree on the other. About 40 metres away I find another half, plain bower. Most likely these are the work of a young male still learning the finer points of bower construction. For the moment I have to be content knowing that a golden bowerbird does live here and that one day he will build a proper bower. But as male golden bowerbirds take about seven years to mature, it may not be any day soon.

Droppings so fresh that the bird should be just ahead of me, eggshells from a hidden nest, songs heard but the birds not seen, at-the-bower whistles but no bowers, an owl's hiding place but no owl, an incomplete nest, an apprentice's bower but no apprentice. Today the forest seems inhabited only by phantoms.

A Bird with Built-in Sunshine

Bulurru, 29 September

A certain mysteriousness is one of the forest's greatest attractions and I am content in the knowledge that many of its mysteries will never be revealed. On the other hand a day like yesterday, full of phantoms, does not sit well. This morning I set out early to try my luck again, especially with the golden bowerbird.

As I walk down to the creek along a different route the omens are good. A carpet snake lies loosely coiled in a low bush at the forest edge, basking in the sun's first warmth. Just inside the forest I find the nest of a grey-headed robin, built, as it so often is, on the prickly stems of a common wait-a-while. I cannot help thinking the birds must have pierced their feet repeatedly while building it. There is a single egg in the nest.

At the platypus pool, without a platypus this morning, I go downstream a little. Instead of climbing the sheer rock bank I walk up a steep spur that leads to the same ridge. This way I approach the incomplete bower from the opposite direction. Maybe, just maybe, a completed structure is on this spur. I have not gone more than thirty metres when I find a pile of sticks placed at the base of a small tree. I pick up the topmost pieces. They come away easily from the others. This means they have been freshly placed. Sticks that have gone through a prolonged wet spell are fused together by a wood-decaying fungus, something that gives a bower a certain amount of rigidity. Not five metres away there is another pile of sticks, this time wedged between a vine and a tree at about chest height. Because of the height and placement of this structure I realise, with mounting excitement, that this is not the work of a young bird. This is a subsidiary bower of the kind many male golden bowerbirds build on the periphery of their main bower. A proper bower *must* be near.

Before I can rush on I am stopped by the sound of pig-like grunting coming from a thicket of young palms. It goes through my mind that this must be one of the feral pigs that has been in the area of late. I am not totally convinced it is a grunting pig for there is an unusual hollow tone to the sound. All my senses fully alert I stand very still by the subsidiary bower. I take more notice of the whistles I have been hearing: single notes rising at the end like a question. I had put them down as yet another call of a shrike-thrush. But then one of the whistlers gets up from the leaf litter and walks up the hill. It is about 25 centimetres tall, striped in black and yellowish white, with a red-brown head. A cassowary chick, no more than a week old! The grunting comes from its father. I still cannot see him but his voice comes nearer. Two more chicks are on the slope below. They now walk towards me, whistling. Male cassowaries, who look after the chicks on their own, can be very aggressive when they judge their family to be threatened. The chicks come closer. I have visions of them seeing me, tall and two legged, as their father and sticking close.

Should I flee or stay? I reason that the noise of my footsteps would only attract the wrath of the adult cassowary. I stay quietly behind a tree. The two chicks, heads held confidently high, march past about three metres away. Just up the slope their father, resplendent with blue and red bare skin on his head and neck and crowned with a moderately sized casque, materialises out of the shrubs and leads his brood up the ridge. Soon all of them are screened from view. Total silence. Are the birds just behind these bushes, the male ready to come rushing out, or have they moved on? I wait. After about ten minutes I climb, as silently as I can, up the spur.

Seventeen paces from where I stood by the subsidiary structure, its yellowish decorations blazing in a shaft of sunlight, stands the bower of a golden bowerbird. A proper bower this time, one large pyramid of sticks

A male cassowary leads his newly hatched chicks.

and a small one joined by a horizontal stick—the display perch. The smaller pile is adorned with pieces of pale yellow beard lichen while at the place where the perch joins the larger pile, yellowish seed capsules, some still containing shiny black seeds, are tucked between the sticks. The bower is three-quarters of the way up the ridge and only about 80 metres from where I found the subsidiary structures yesterday.

A shadow passes overhead. I look up just in time to see a flash of golden-yellow streak down the hill. I creep slowly to within about ten metres of the bower and hide among the foliage.

The first I know that the bird is near is when I hear his strange frog-like sounds. They are very difficult to describe—a sort of pulsating rattle that has been likened to the ratcheting noise of a fishing reel. The sound comes from behind me, then from my left.

He lands silently on his display perch in a dazzling flurry of golden-yellow. Fluttering down to the ground he retrieves a seed capsule that had become dislodged and places it carefully among the others on the bower. Minutes pass while the bird thoroughly inspects his bower. The sun has moved and the bower, and the bird, are in deep shade. Yet as he moves about, showing the patches of brighter yellow on top of his head, his underside and his tail, it appears as if his plumage is highlighted by dapples of sunlight. The structure of the feathers, their colours and pattern conspire to give the impression that the bird is bathed in light even on the dullest day. The golden bowerbird carries his own sunshine with him.

A grey-headed robin broods its egg in a nest built among the spines of a wait-a-while vine.

The sooty owl, the 'phantom' that produced the pellets. This bird is in immature plumage. In adults the underside and face are almost white.

Suddenly the bird stops his relaxed fiddling with sticks and ornaments. He is fully alert and calls a harsh 'kek, kek, kek'. Leaning back on his perch he half opens his wings, exposing more brilliant yellow. A plain brown bowerbird, presumably a female golden, lands in a small bush near the bower. The male picks up some lichen and holds it in his bill. The female takes off and he follows her, still carrying the lichen. Moments later I hear the male very close. He sings as if in a whisper, imitating figbirds, drongos and king parrots. The female returns, landing on the central perch. The male flies at her with his back towards me. As he passes over the bower he spreads his tail and flashes its bright yellow outer feathers. The female hurries off. Twice more she comes and twice more he seems to chase her away. The next time the female lands in front of the bower and the male bows to her with a *Melicope* capsule in his bill and spreads his wings slightly. Is he courting a female, as seems to be the case when he postures with open wings and ornament in his bill, or is he chasing her off? Or maybe it is not a female at all but an immature male who to human eyes is indistinguishable from the female? I cannot tell.

So many of Australia's plants and animals, like the *Austrobaileya* vine that grows beside the bower and the platypus swimming in the creek below, are ancient and primitive. But the very opposite is true of the bowerbirds and their close relatives the birds of paradise, represented here by the riflebird. Among the birds they are the most advanced. In terms of courtship displays and rituals they are at the pinnacle of evolutionary development.

This golden bird that lives high in the canopy and that ordinarily you rarely see, yet is bold and unafraid at the bower, embodies all the beauty, romance and mystery of these forests.

I continue along the ridge and pass the hollow tree where the owl roosts and learn once again never to jump to conclusions about the forest's wildlife. I approach the tree carefully, looking up this time and not at the ground looking for pellets. I hear just the faintest sound of wings brushing some leaves and see an owl fly from the hollow to a small low tree only a few metres away. It is not a rufous owl. The eyes staring down at me are not yellow but huge and black and belong to a sooty owl, the largest I have ever seen. She—I surmise it is a female as these are larger than males—is barely 'sooty'. Her stout feathered legs are white and much of her underside a pale dirty grey. Her wings are a silvery mid-grey covered with white spots.

This morning's golden apparition stays in my head and imagination. I realise there is much more to this extraordinary bird than I saw today. But that is not something you discover during a few hours or even a whole season. To unravel something of the bird's life history needs a 13-year study, such as my friends Cliff and Dawn Frith have carried out. The Friths live in rainforest a few kilometres away over the ridges. I arrange to meet them to watch and talk about the golden bowerbirds.

A Passion for Bowerbirds

Prionodura, 4 October

The golden bowerbird is known to science as *Prionodura newtoniana*. Cliff and Dawn Frith have named their rainforest home Prionodura after the bird that has been a preoccupation of their lives for many years.

The day's first glimmer tints the sky behind Mount Bartle Frere as I enter the Friths' forest. Cliff and I set off to visit the bower of a golden bowerbird. It is only just light enough for us to see the trail ahead. Birds sing everywhere. The exuberant dawn chorus on these fine spring mornings makes the whole forest ring with sound. Cliff easily identifies the various songs. He points out the ascending cadence of the grey fantail's sweet little song. The closely related rufous fantail sings in slightly higher pitched notes in a descending scale. A Lewin's honeyeater gives a burst of its staccato call. Cliff draws my attention to the warblings of several silvereyes.

The sun finally rises and illuminates the trees' topmost branches with a yellow light. The sky is clear and blue. Two groups of chowchillas call close to us. In one group two birds stand toe to toe on a mossy log and sing at each other in rhythmic voices of incredible volume. It seems that they must deafen each other. In quick succession Cliff points out the songs of a golden whistler, the trill of a fan-tailed cuckoo, the thin little reels of brown warblers, the bubbling voice of a white-throated treecreeper, the

A male golden bowerbird on his bower.

sharp notes of a large-billed scrubwren, the squeaks and churrs of a mountain thornbill and the pleasing lazy whistle of a black-faced flycatcher. Brown and purple-crowned pigeons call. Cliff stops and listens intently. 'Can you hear the fern wren?' he asks. I can just make out the high pitched rising call, so high pitched that it is only just within my hearing range.

I can clearly hear, however, an insistent ascending trill, a song that comes closer and closer. Suddenly a small bright yellow bird with a broad black bill comes into view. He sings and sings, his throat fluffed, his tail cocked to the vertical like a wren's, his wings slightly drooped. We have seen singing yellow-breasted boatbills before, but never one giving such a sustained song and never in that particular posture. We can only guess at its meaning—is the bird's nest nearby, is he courting a female which is hidden from us?

A male yellow-breasted boatbill singing.

The chorus swells and continues as many more species join in. For nearly an hour we are enveloped by sound.

We can hear the golden bowerbird's ratcheting call well before we come to his bower. When we are close we see the bird, a splash of bright yellow, adjust some ornaments and then fly off. We quickly put up the hide Cliff brought. He will use it over the coming months to continue his and Dawn's observations of this species.

We walk on. We pause at a small rock-lined pool where whirligig beetles skate over the water surface, then climb a steep slope till we come to a broad flat ridge top. Tall buttressed trees are widely spaced. Among them are elegantly shaped shrubs of the yellow wait-a-while as yet without their lines of hooks. Huge vines snake up to the canopy. Yellow fungi glow on a rotting log. The edge of this miniature plateau drops sharply down to a creek we can hear rushing along. The sun rises over the opposite ridge and its light floods through the trees. We settle on two conveniently placed rocks and take in all this dazzlement before talking about rainforests and bowerbirds.

Cliff is dressed in drab green T-shirt and trousers. He is in his early forties and of medium build. His hair is dark brown while his closely cropped beard is streaked with grey. His eyes too are dark brown and sparkle and glitter when he becomes enthused about the forest and the birds, which is often. Despite his intense feelings about both subjects, he does not pursue them with a humourless single mindedness. On the contrary, he has the capacity to see humour in most situations.

Cliff's origins could not have been further removed, in distance and in spirit, from north Queensland's tropics. He was born in south London, one of ten children. None of his large family is remotely interested in natural history. His brothers and sisters look upon Cliff as an oddity and when they come to visit Prionodura they are a little apprehensive about the surrounding wild and untamed environment.

From an early age Cliff knew that city life was not for him. He was always engrossed in natural history books, particularly those about birds, and whenever he could he rushed out to the woods.

Dawn's background is very different. She has no brothers and sisters. She grew up in the relative tranquillity of the Gloucestershire countryside with traditions of horse riding and nature walks. But like Cliff she was driven by an interest in natural history, an interest neither shared nor understood by her parents. Dawn pursued an academic path that led to a PhD degree in marine invertebrates and then to teaching.

Cliff, in the meantime, had seized an opportunity to go to, if not the rainforest, at least the tropics. It was one step closer to the place that from earliest childhood had been to him the ultimate habitat, a romantic place full of exciting things to see and with endless possibilities for new discoveries. In 1968, aged 18, Cliff joined the British Museum of Natural History Harold Hall Expedition to the Kimberleys and Arnhem Land in northern Australia to study birds. While he had been captivated by bowerbirds long before this, he had never seen one in the wild. Cliff tells me that he can still clearly remember finding his first bower. It was of a great bowerbird. He lay on his stomach a few metres from the bower, waiting for the male bird to arrive, which it did. Cliff took some pictures with rather primitive camera equipment, but it was enough to cement a lifelong association with these birds.

The original stimulus to seek a life among bowerbirds in Australia's tropical rainforest came from two books. 'David Snow's *The Web of Adaptation* really intrigued me,' explains Cliff, 'especially his work on fruit-eating birds in tropical rainforest. The bottom line of this study, from his theoretical standpoint, was that there are promiscuous males in species like cotingas, manakins and bowerbirds because of the richness and diversity of fruits in the rainforest. This means a superabundance of food which gives the male birds time to display and the females the ability to raise the young unaided. So it was this connection between fruit diversity and the ways of life of the bowerbirds that meant we had to go to the rainforest.

'The other book is A. J. Marshall's *Bower-birds: Their Displays and Breeding Cycles.* Under the species accounts of both the golden and tooth-billed bowerbirds he mentions Seaview Range. I wrote in my copy, many years ago, "sounds a wonderful place", because the implication was that you were in the rainforest where these birds occur, but you could see the ocean. That always stuck in my mind. Then when we got to Australia a colleague who had done some basic work on rainforest birds said, "Oh, the place you should go to is Paluma in the hills just north of Townsville". I didn't at that time realise that Paluma is on the Seaview Range. That was quite a coincidence. We took the advice and drove from Sydney to Paluma in a Land Rover towing a large caravan. Paluma turned out to be a very pleasant community and all that we had hoped it would be. We lived there for twelve years. Dawn and I always said we would not leave Paluma unless we found something better, which eventually we did.

'As soon as we arrived in Paluma we began intensive—when we look back on it we sometimes think too intensive—research. We spent three years literally non-stop, from early in the morning till late at night, day after day, accumulating information on the bowerbirds, the golden, the tooth-billed and the spotted catbird. The catbird is also a bowerbird, but it doesn't build a bower or a court. To a lesser extent we also studied the satin bowerbird.

'Except for the satin there was really nothing known about these bowerbirds other than a few anecdotal publications which, as it turned out, were misleading.'

I ask Cliff what, in summary, he and Dawn discovered about those extraordinary birds. He tells me they described the entire nesting biology of all three species in great detail, and that they have records of the birds' longevity. In the case of the golden and tooth-bill, they gathered a great deal of information on bower and court construction, courtship behaviour, the interactions between males of different ages, how males become bower owners and much more. As a means to better understand the birds' ecology they also undertook broad studies of the leaf litter animals, the insect populations and the fruiting trees—the birds' three main food sources.

Bowers are such amazing structures, elaborate and richly decorated and unlike anything else in the animal kingdom, that much has been written about them. There has been a great deal of speculation and theorising about just what bowers represent, why they exist and take the form they do. I ask Cliff what his findings are.

'We've looked a great deal at this. We conclude, as do a number of other workers, that bowers symbolise male sexuality and clearly indicate to females the genetic fitness of males. That is the essence, but there are so many subtleties related to that. Some bowerbirds paint their bowers and others decorate them with really remarkable things. One generalisation that we've come up with and managed to prove is that bower decorations tend to be rare items, and that they are rare items for a very good reason. If they weren't rare they wouldn't truly symbolise the male's fitness. Perhaps the most extreme example of that is a discovery we made in Papua New Guinea recently. We found that male Archbold's bowerbirds use the plumes of King of Saxony birds of paradise to decorate their bowers. These are very, very rare.

'In the case of the golden bowerbird, who uses beard lichen and the seed capsules of *Melicope* shrubs as decorations, you might argue that beard lichen is not rare. But this tends to grow high in the canopy, which is not a place the birds spend a great deal of time. *Melicope* is a shrub found at the edge of the forest and may be fairly easy to find now. But before logging and clearing and other artificial damage to the forest, I think they were rare items.'

Some early ornithologists were so impressed by the combination of the bowerbirds' building and decorating abilities and their vocal prowess that they placed them at the top of the evolutionary scale among birds. Some scientists even went so far as to suggest they occupied the same place among birds as monkeys and apes do among mammals. I ask Cliff about his views on this.

'In a purely evolutionary sense I don't think that is true. But in another

A male golden bowerbird perched on a wait-a-while vine, carries beard lichen to decorate his bower.

sense I can understand that statement, and I must agree that bowerbirds are extremely advanced in particular ways. Behaviourally they are very, very complex, more complex and more sophisticated than other birds. But to say that they are evolutionarily the most advanced I think is misleading. It implies intelligence and I think that crows, for example, are more intelligent than bowerbirds. It's the bowerbirds' artistic abilities, and the subtleties in the way they build and decorate their bowers, that leads to the perception of intelligence, which I think is wrong. But artistic ability, now that's a different matter. Several authors are currently suggesting that maybe bowerbirds show the first real steps towards an artistic culture, and I think there's a lot of evidence to support that. We now know, for example, that bower building is not truly instinctive. Males must learn how to construct the right kind of bowers so that females will accept them. In that respect it is truly cultural.'

I tell Cliff of the things I have seen at the golden's bower at Bulurru and the startling brilliance of the male in the subdued light of the forest. Cliff asks me if I have ever seen the male's hovering display.

'It's the most theatrical display and shows the bird's unique plumage to greatest effect. It is brightly coloured on its underside, which most bowerbirds are not. During the display the male flies, with slow, butterfly-like wingbeats, around the bower. At a few points, his bill almost touching the trunk of some sapling, he'll just hover. At the same time he spreads his tail, flicking the brilliant outer tail feathers from beneath the dull central ones. I remember once seeing two golden males perform this flight display simultaneously at a bower deep in the forest without apparent aggression towards each other. It was quite staggering to see this intensity of colour, like flames, in the gloom of the undergrowth. I don't mind admitting I was shaking with excitement when I saw that.'

The Friths are not funded by any institution or government grant. They have become world authorities on bowerbirds and birds of paradise entirely through their own efforts. During their intensive studies at Paluma they lived on their savings and the income from Cliff's wildlife photography.

So often when people intensively study certain groups of plants or animals they become rather one-eyed about the forest as a whole. But the Friths appreciate and have feelings for all rainforest life. No doubt photography has helped to maintain the wide base of their interest in and enjoyment of the forest.

I ask Cliff what his and Dawn's plans are for the future, what else they would like to achieve.

'There are many things. Unfortunately I'm not as broad an ecologist as I would like to be. So I'm never going to understand the system as a whole. I don't think many people will achieve anything close to that. My primary interest is in the birds and how they relate to the ecosystem. Obviously one has personal little ambitions and my obsession is bowerbirds and birds of paradise. I'd like to learn as much about them as I can in my lifetime. Much remains to be learned, Dawn and I have only scratched the surface. We can describe the fundamental nesting biology, the behaviour of the males at their bowers and the relationships between males. But there is still an enormous amount of work to be done to understand the origins of these behaviours, the ecological restraints on the females at the nest, the genetic relationships between female offspring and the male to whose bower they may eventually go. Do females nest close to the males who fertilise them? If they do, does that imply a possible future social relationship between any given male and his offspring? And much more. There are several lifetimes of work to be done to understand these things.

'But all these ambitions pale into insignificance compared to our goal of trying to make people, all people, more aware of the need to preserve every square inch of rainforest that is left on earth. I mean, everything else inevitably pales into insignificance compared to that. We like to think that everything we do is aimed at that ultimate goal. There are many ways of achieving it. You can stand in front of the bulldozers or you can make people, in the broadest possible sense, more aware of what we stand to lose by destroying rainforest. In the long run the latter is, I think, more effective.

'You know,' Cliff confides as we walk slowly back to the house, 'you can take a walk one particular morning, like today, and the beauty of the light and the vegetation is just overwhelming. There are moments every day with different animals and different plants that astound and touch you. When I'm walking in the forest I'm always looking for birds' nests. I well remember the first time I saw what I thought was a nest of a bird at the base of a tree. When I touched it, out came a musky rat-kangaroo. That was just amazing, to think of a tiny primitive kangaroo that builds a domed nest out of leaves. You contemplate that for the next few days and eventually absorb that remarkable piece of evolution and try to assimilate that into everything else that you see happening on the forest floor. Every day of every month of every year is a constant accumulation that adds to the wonder that is tropical rainforest.

'The cassowary was one of the birds I had dreamed of meeting since my earliest childhood: a flightless bird as large as yourself! My first encounter was unforgettable. It was a cold winter's day at Paluma. At that time Dawn always wore a blue woollen beanie. So when I saw this apparently blue hat, at about Dawn's head-height, deep in our study area I thought it was Dawn approaching. Only when I was very close did I suddenly realise that this was a cassowary. I just froze, confronted by this staggering bird in its own habitat. I mean, we were here to study bowerbirds because they eat fruit and have this special relationship with the forest as a result of that. But here was an enormous bird that relies almost entirely on fruit and supports a body as large as my own. If there was a single experience in the rainforest that was utterly humbling, that was it.'

Vocal Gymnastics

Bulurru and Prionodura, 10 October

For some days I have been watching and listening to a tooth-billed bowerbird at his court. I thought him to be very close for I could hear him so clearly. But I had to walk down the gully to the west of the house, to the top of the opposite slope and then a short distance down the other side before I found his display area. He had meticulously cleared an oval space, almost two metres across at the widest part, of all leaves, sticks and twigs. It looked as though it had been swept clean. A few saplings grew in and around this oval. The bird had brought fresh green leaves to this clearing and placed them upside down, exposing their paler undersides. I recognised the leaves of the umbrella tree, the Davidson's plum, the white aspen, the rusty laurel, and one from a ginger plant that was 55 centimetres long, more than twice the length of the bird. Old, curled but still green leaves lay discarded around the periphery of the court.

I want to have a really close look at the tooth-bill at his court so I have put up a hide about three metres away. This morning I am in the hide before sunrise, before the court's owner arrives. I have barely settled on my folding stool and made a small peephole in the hide's cloth, when I hear him approach, hopping over the dry leaves. Apart from the faint rustling he arrives as a silent dark shadow in the twilit forest. He nearly trips over the large leaf he carries lengthways in his beak. Carefully he places it upside down among the other green leaves. He looks at it for a while like a patience player might at a newly turned over playing card. The rest of the court is also closely scrutinised. One after another he picks up yellow and brown leaves that have fallen during the night, takes them to the edge of his court and tosses them aside with a flick of his head.

With the court now in order, he is ready to begin the day's performance. He hops onto the curving stem of a vine, almost 50 centimetres above the ground to the right of the court and looks all around with his bright, almost black and very alert eyes. I hold my breath, hoping he will not notice my eye pressed to the hole in the hide. He stares hard in my direction for a few moments then looks away. Slowly he tunes his voice with sounds like 'clee' or 'chee', 'kerek', 'kerek-eh' and a few scratchy notes. After a while he adds clear loud whistles and soft whisperings, then gradually he goes into full song, a masterful blend of his own varied notes and imitations of a great many of the other voices of the forest. It is not an unending torrent of sound but rather a collection of phrases alternating with brief pauses. During the pauses the bird looks alertly all around and seems to be listening. He obviously hears the king parrots land in the trees above us. The parrots converse in a series of clear whistles and calls of 'chack, chack'. Immediately the tooth-bill weaves perfect, at least to my ears, imitations of these notes into his song.

On this fresh clear morning other birds are in fine voice. Golden whistlers call close to the court and a band of chowchillas start a boisterous chorus. Passages from these disparate songs soon appear in the tooth-bill's repertoire.

After 45 minutes of almost constant song, it suddenly stops. The sounds were so close and so loud that the silence comes as a relief. Twelve minutes later the tooth-bill returns with another fresh leaf and places it among the others. For a few more minutes he rearranges the leaves till they are placed to his satisfaction.

This time he hops onto a low branch even closer to me. Again he first tunes up and then gradually launches into his full-throated song. As he

A male tooth-billed bowerbird sings at his court.

sings he opens his beak wide, showing his white palate and the pale rim of his lower mandible. His pale throat feathers and to a lesser degree those on his cheeks are fluffed out and vibrate as he works his voice box.

He now imitates a range of other birds: the fluting notes of a black butcherbird, the rat-tat-tat of a Lewin's honeyeater, the scratchy rollicking song of a bridled honeyeater, calls of the golden whistler, the subdued whisperings of a scrubwren. Not only does he mimic the sounds but also their varying volumes. The effect is such that if you just listen to the bird sing, without seeing him, you have the impression that the different songs come from different places; that you hear a number of birds, some close and others further away. But all the sounds are made by just this one bird, sitting in the one place. At other times I have heard tooth-bills mimic a grey goshawk, a grey-headed robin, a crested hawk, a drongo, a figbird, a black-faced flycatcher and once the screech of a flying-fox.

Apart from the rather clear and melodious sounds, he sometimes produces a harsh strange noise; a wheezing kind of screech followed by the penetrating crackling sounds made by certain cicadas. But even that is not the end of his vocal gymnastics. At first I think a Lewin's honeyeater has joined him in a duet, but I soon realise that the tooth-bill utters the wheezing screeches and the honeyeater's song himself and both at the same time. The tooth-bill has this uncanny ability because its syrinx (the part of its throat that produces the sounds) is divided, enabling the bird to sing with two voices at once.

Eventually the tooth-bill moves away from his singing perch at the court. For the next hour he sings from a number of places higher in the trees, broadcasting his song over a greater area.

My limbs and back stiff from sitting in the cramped hide, I stumble out and walk back to the house. All the while the tooth-bill continues to sing. From a distance, however, I cannot tell that he sings with two voices, nor can I detect different pitches for the different mimicked songs. From here his is just another voice in the forest. A voice that hardly rests all day.

When out walking, when you come across a tooth-bill's court, you think how marvellous, how clever. But when you sit in a hide by a bird's court for a morning, when you see the meticulous way he keeps this clear, when you observe how carefully he collects and places the green leaves, when you notice how bright-eyed and alert he is, when you are witness to his extraordinary vocal powers, then your feelings for this bird run much deeper; you are overcome with wonder, with disbelief almost, at his methodicalness and sense of purpose.

The purpose of the court and the singing of this plain bird is to attract a mate. When he has attracted her he has one more trick up his wing, or his cheeks more accurately, to impress her further. During the many days I have spent watching male tooth-bills at their courts I have not yet seen a female near any of them. What happens when she does visit? And does the female build the nest and care for the young on her own, as is the case with other bowerbirds? Do males come back to the same place every year to build their courts? Do males who have courts close together interact?

To seek answers to these and other questions I return to the Friths. Dawn did most of the actual field work on the tooth-bill. For many hours of many days Dawn sat in hides to observe the birds' behaviour at their courts. She recorded their songs and watched their displays and rituals. She also measured courts, banded the birds so she could identify individuals, studied feeding habits, and followed radio-tagged birds.

Tooth-bills are singing near the house when I arrive, so Dawn and I decide we may as well sit in the forest, near a court, while we talk about these birds. Fit and tanned after a recent trip to study bowerbirds and birds of paradise in New Guinea, Dawn moves easily and swiftly through the undergrowth. She is obviously very much at home in the rainforest. We sit on a log within sight of a court where a male sings, and Dawn patiently and methodically answers my questions.

My burning question is, what happens when a female tooth-bill arrives at a male's court? Dawn relates some of her experiences:

'During my observations I noticed that on each court there is a favourite sapling, usually larger than the others. Often this thin tree has a small buttress. The bird clears the space around this sapling meticulously. He doesn't put leaves close to it; it is his display tree. When he notices a female nearby he dives behind that tree, hiding from her. Now and again he'll peek at her. If she comes near the court the male goes into this quite incredible display. First of all he changes to a very, very quiet subsong, almost like whispering, which consists mostly of mimicked notes. All the while he moves about behind the tree, peeking and watching to see what the female is doing. His voice becomes quieter and quieter. When the female is close to his tree the male suddenly shoots out from behind it. Flicking his wings out he rushes towards her. At the same time the feathers under his throat, which are almost white, puff out. So he suddenly appears, whispering, with a white puff-ball on either side of his bill. If the female stays, the male then goes into a kind of bouncing display.

'I've probably seen this sequence of events forty or even fifty times, but never have I seen them mating. Always the display ended in the two birds flying away. Cliff and I were very excited when we first saw these courtship displays. They had never been described before. Even now we haven't published these observations. We have so much unpublished data on both the golden and the tooth-bill that we've reached the point

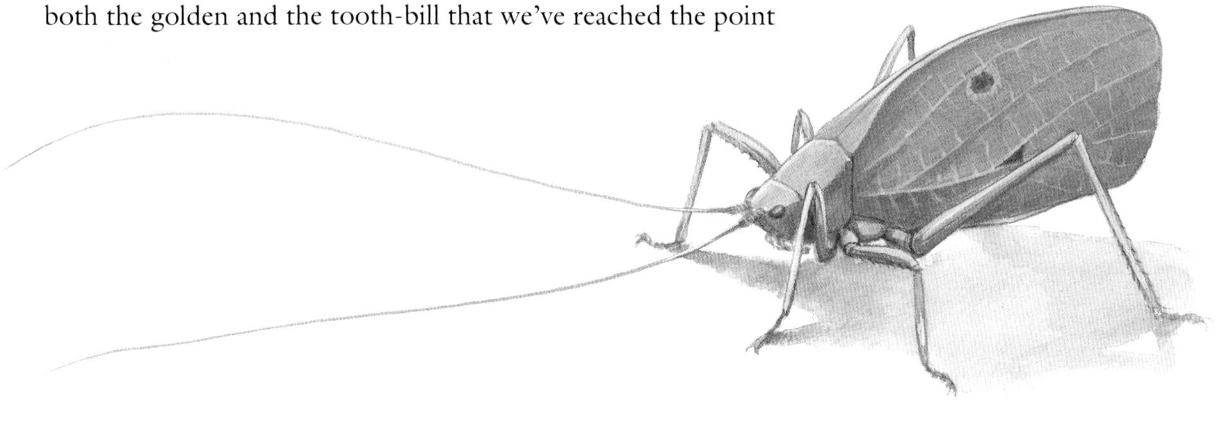

Many rainforest animals from lizards and frogs to insects, are almost impossible to see because of their camouflage. This katydid resembles a leaf complete with veins and such imperfections as fungus spots.

where we have to stop the other things we're doing and write it all up. If we sat down tomorrow, which is an impossible thing to do, and both of us worked full time, it would take us about eighteen months to two years to do.'

Another question at the forefront of my mind is about the bird's name. Both males and females have a notch in their beaks, the so-called 'tooth'. I ask Dawn if she and Cliff found that this 'tooth' was exclusively used by the male to bite or chew off green leaves for his court. Dawn says that is probably not the case, because other species of birds, who do not have a 'tooth', can also pull off leaves. Also, she points out, both male and female tooth-bills have this notch and the females neither build nor decorate courts. The special tooth-like part of the bill is more likely to have something to do with the birds' diet at certain times of year when they eat leaves, not fruit.

I ask Dawn if their study turned up any surprises.

'Yes, there was one major surprise and that was that some males are thieves. They take the easy way out and steal leaves from other courts. That was an interesting discovery as no one had suspected that. I discovered it while looking at the rate of leaf turn-over. I had selected nine closely spaced courts. Every week I labelled each green leaf on the courts with the date and the number of the court. The following week I'd be back to note how many new leaves there were and label those. This way I was able to work out how long a leaf stayed on a court. I found out that ninety per cent of the leaves had withered to the stage of being discarded after fourteen days.

'Then leaves I'd marked on one court started appearing on others. Because the birds were colour-banded we knew each individually and we worked out who the thieves were. One was the greatest thief. He stole from seven adjacent courts but none of his leaves were stolen. We're not sure yet what this means. It could well be that the thief is the dominant male of those nine courts. But we can't say yet, we'd have to analyse our data further.

'I've become very fond of the tooth-bills over the years. I suppose it's because I know so much about them. Some birds we have watched on courts in the same place in the forest for thirteen consecutive seasons. I enjoyed learning about them, sitting in hides and just watching them for hours at a time. What is so exciting when you colour-band birds [two plastic bands, each of a different colour, are placed on the birds' legs] is that you get to know individuals. They sort of become friends. You learn what white/blue is doing and where red/green has gone.'

Paluma, where the Friths did their studies, is nearly as wet as our area here. But it is at a considerably higher elevation so that it is much cooler, with longer periods of mist. I ask Dawn if she minded all this rain and getting wet. She says:

'I don't mind too much. If we were making our three-hour observations from the hide and it began to rain we'd sit through it. I remember once while doing some observations at a catbird's nest it started to thunder and rain. The parent bird came in and just sat tight on the nest. And I just sat in the hide. She remained there for over an hour, in the pouring rain, getting absolutely saturated protecting her chicks.'

As Dawn and I walk back to the house I am once again struck by how different she and Cliff are despite a shared interest in birds and tropical rainforest. Cliff, dark-eyed and dark-haired, is more intense. He is very much interested in the theoretical and philosophical aspects of their work. Dawn, with long blonde hair, high cheekbones and light blue eyes, is more easygoing and also more methodical, with a concern about close observations and the nuts and bolts of just exactly how things work in ecological terms. They complement each other brilliantly. Between them they have increased and spread our knowledge of the rainforest through their studies and their books, which somehow they find the time to publish themselves. Long may they continue.

Bolwarra Weevils

BULURRU, 15 OCTOBER

At all times of the year the forest is full of scents. Not all of them are sweet or even pleasant. A small shrub growing close to the house looks very attractive, has glossy dark leaves and neat rows of red fruit. The combination of leaf and fruit has given rise to one of its names, native coffee bush. It is also known as bridal bush, though I cannot imagine why. When you brush against it or crush some of its leaves, the innocent looking plant gives off a terrible smell: the smell of gas known as methyl mercaptan, which is also present in human faeces. It is more accurately known as the fart bush.

Hairy rosewoods, medium-sized trees, are in fruit at the moment and give off a similar odour. The very special and primitive vine which scientists call *Austrobaileya scandens* has just finished flowering. The flowers had a strong smell of ammonia and urine. During the wet season various fungi known as stinkhorns and stinkstars emerge from the leaf litter. They are covered in slime that smells of rotting meat and attracts blowflies. The slime contains the spores which are dispersed by the flies. Sometimes when out in the forest an awful smell wafts across the trail. It is so strong that you are convinced you have stepped in something nasty. But so far I have not been able to track this smell down. Luckily these odours are rare, and the sweet and fruity

Bolwarra flowers with their weevils.

scents—musk and lemon, gardenia and violets—are much more common.

At daybreak I pick up an exceptionally strong and agreeable perfume; a combination of musk and some sweet fruit. The bolwarra, a scrambling bush growing about 30 metres away, must have come into flower. It is one of the most primitive flowering plants. For some weeks I have been watching the flower buds grow and swell. This morning three buds have pushed off their caps and burst into flower, spreading a profusion of what look like pure white petals, shining as though they had been enamelled. But things are not always what they seem, even among flowers. The immaculate 'petals' are in fact sterile stamens that have assumed the shape and function of true petals.

I am not the only one attracted by the heavy perfume. Tiny weevils only a few millimetres long land on the flowers and burrow into their depths. There are so many that all the 'petals' heave and move. I pick one flower and count 46 weevils buried within it. Another carries an even bigger load, 60 or 70 of the beetles at least. These minute insects are the bolwarra's only pollinators and the scent is there to attract them and only them. No other insects have ever been observed on or in the flowers.

Many rainforest trees, such as the satinashes, spread their scent and their nectar for all and sundry. Their flowers may be pollinated by beetles, moths, butterflies, thrips, wasps, bees, ants, flies or even cockroaches among the insects. Birds, flying-foxes or possums may perform the task equally well. Usually it is done by a combination of all or some of these animals. Other plants, like the bolwarra, are more specialised in just who pollinates them.

Not a great deal is known about the pollination of Australia's rainforest plants. As it happens the bolwarra was the subject of one of the first pollination studies in the country. It was discovered as long ago as 1860 that the tiny weevils are its sole pollinators.

By crawling all through the flowers the multitude of beetles transfer the pollen from the fertile anthers, the male parts of the flower, to the stigmas, the female parts. This fertilises the flower which then develops into a fruit. The plant now has no further use for either the pollen bearing stamens or the infertile ones that act as petals.

By the end of the day the weevils have laid their eggs on these floral parts, which are in fact fused together into a single mass. This mass has become soiled and brownish in the course of the day and at nightfall falls to the ground; what had been a brilliant flower this morning is now a smooth flat disc. During the year this will develop into a top-shaped yellowish fruit. When ripe it will be full of seeds suspended in a clear pulp whose scent, though different from the flower, is also attractive. The fruit is edible and is also known as the native guava.

And what of the weevils? Their eggs will hatch in a few days and the larvae will feed on the discarded stamens. In a little over two weeks a new generation of weevils will be ready to track down the scent of more bolwarra flowers. But there is a hitch in this seemingly perfect arrangement: by the time the weevils emerge this bolwarra plant and all the others nearby will have finished flowering. There will be no more blossoms till this time next year. How will the insects survive in the meantime? On other flowers? No one knows. But what is known is that wherever bolwarra flowers open, these weevils, and only these, invariably appear to pollinate them. When a bolwarra flower is picked and taken to rainforest where no others are known to occur, the weevils come flying in to the scent within the hour.

Of White Possums and Wooden Castles

MOUNT LEWIS, 19–20 OCTOBER

Late in the afternoon we pull up at a small clearing among trees at the junction of two creeks about halfway up the mountain. As soon as Bill, Wendy and I step out of the car we realise that this is very different rainforest from the one we are used to. There are familiar plants: blue quandong, tropical quandong, Murray's laurel, hard alder, brush mahogany, orania palm, piper vine, gardenia. But some are very different. Wendy recognises a briar silky oak, a rose silky oak, a rose alder and others we do not know. Vines are fewer but shrubs and saplings more numerous. Apart from the different plants there is a different look and feel about the place, especially along the creek. The pebbles, rocks and boulders are a rough textured granite. At the bottom of the pools and on exposed spits there is coarse white sand where flecks of mica glint like gold in the sun.

Mount Lewis crayfish.

The ridges slope gently down to the stream. At Bulurru the rocks are smooth basalt. Pools and exposed banks are lined with red soil which readily turns to mud. Slopes are steep.

Between the rainforest of Mount Lewis, the adjoining uplands of the Mount Carbine and Mount Windsor Tablelands and that of our region about 130 kilometres to the south, there is a narrow corridor of dry country. The two regions have been isolated from each other for some time and their rainforests are quite different. A whole array of plants, insects, frogs and other organisms are found only on these northern uplands and nowhere else. Some are exclusive just to Mount Lewis. It is to see these endemics that we have come here.

The waterholes are so clear that the water is almost invisible. Enormous tadpoles of the northern barred frogs swim among the fallen leaves that have sunk to the bottom. The leaves form a random mosaic of browns, yellows, greens and reds. Bill suddenly exclaims: 'There's one!' I carefully scan the bottom of the pool. A small corner of the mosaic has a slightly different shape and subtly different colours. Slowly these resolve themselves into a large crayfish, a brownish individual with trimmings of blue and red. Then we see another, and a third. This is the Mount Lewis crayfish, one of the endemics.

The crayfish are feeding. Their large colourful claws are held out straight in front of them. These are only for ripping and pulling in combat. Close to their mouths, each has a pair of small white claws; with these they rapidly gather minute particles of rotting leaves or perhaps the algae growing on them. We thought that the tadpoles swimming all around the crustaceans might occasionally provide them with a meal. But the incipient frogs swim all around the crayfish, even between their claws, without eliciting any kind of aggression.

Beside the creek stands a tall and straight tree. Its lower branches sweep down almost to the ground. Sunlight slanting through the forest catches the tip of one of the branches, illuminating oval-shaped translucent red fruit about five centimetres long. The tree's new shoots are covered in fine rusty brown hairs which in mature leaves are retained along the veins on the underside. Hence the name rusty carabeen. It is found only in the Mount Lewis region. Bill cuts one of the fibrous fruits in half. The flesh smells like ripe watermelon.

Close beside it stands a small sapling of a briar silky oak. The sun strikes one of the 30-centimetre unfolding leaves. Its colour and texture are quite startling. At the tips the leaves are rose-pink with a purple sheen. They feel like velvet but look like satin. Fine soft hairs create these effects. Down each leaf's central vein there are no hairs and here the colour is vivid yellow-green grading into bronze-green.

The briar silky oak is a special favourite, not only because of its new leaves, but because it provides an important key in understanding the evolution of plants on the continent as a whole. Fossil leaves perhaps as much as 50 million years old very closely allied to the briar silky oak have been found in Anglesea, not far from Melbourne and close to the

A stream on Mount Lewis. The boulders are granite and the pools the home of the Mount Lewis crayfish. On the right an orania palm leans over the water.

southernmost point on mainland Australia.

At dusk we make camp under quandong and silky oak trees. A pair of yellow-throated scrubwrens hop fearlessly between and all around us as we pitch tents and roll out swags. We discuss what we might see as the night air cools under a half moon and small bats weave in and out of the clearing.

About an hour after dark we set off to look for two more endemics: two kinds of possum. these are not different species to those on the Atherton Tableland, but striking colour varieties. As we drive slowly along I use the spotlight from the car. After a few kilometres I pick up red eyeshine. The light beam captures a Herbert River ringtail. We can just make out that it is different from the black and white ones we see at home, but it is too high up in a pink ash to see clearly.

At the highest point on the road we park the car. We must be close to the mountain's 1238-metre summit. We can make out the shapes of dark, humped ridges in the clear moonlight. An unexpectedly cold and blustery wind tears down their slopes, whipping and tossing the tree tops.

Systematically we scan the trees with our spotlights. We pick up the eyeshine of a total of six possums. Four of them are Herbert River ringtails some distance away. The other two are lemuroid ringtail possums. They are much closer, and the same colour as the ones at home. We walk down the road which dips into a sheltered valley. Tall palms with slender trunks come snaking out of the undergrowth below us and continue on many metres over our heads. Other palms, a kind we have not seen before, are smaller and crowd along the roadside. Bill spots a Herbert River ringtail about eight metres up in a small tree. She calmly looks down at us while reaching out for another leaf. Pulling it towards herself the possum resumes her meal. We can only see her white underside, which has a bulging pouch.

After about 45 minutes we turn back; our batteries have only a limited life. Immediately we see another Herbert River ringtail. It sits in the lowest part of a vine slung between two trees and gently rocks back and forth on this natural swing. On the opposite side of the road another, in the topmost branches of a large tree, is being tossed about by the wind.

Mount Lewis stag beetle.

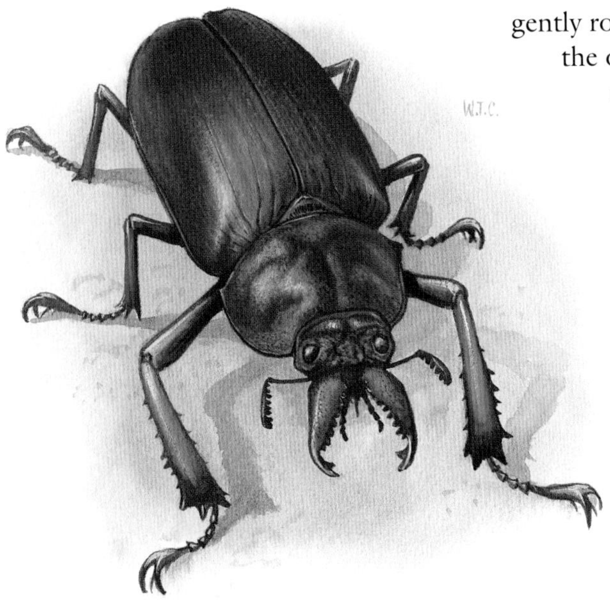

I look for possums deep in the forest. Bill searches both far and near. He stops and simply says, 'Look here.' We are face to face with a Herbert River possum about two metres up a pink ash. Now we can see its pattern clearly: the caramel coloured head and shoulders, the greyish back and flanks, the bare purplish nose. He sits there frozen as we admire his roman-nosed profile with his upper incisor teeth protruding slightly. I briefly touch his soft fur. This colour variety has recently been given a name of its own. It is now the Daintree

Lemuroid ringtail possums have binocular vision and are skilled jumpers and leapers.

The white lemuroid ringtail possum.

ringtail possum and may well turn out to be a different species.

Our batteries are flat by the time we get back to the car, but we still have not seen the other special possum. I volunteer to look for it while sitting on the roofrack with the light plugged into the car battery. Bill drives along at a fast walking pace while I continue to search the trees. I spot more Daintree ringtails. We see eighteen in all, and several brown lemuroids. We are a little dejected as we turn around at the end of the road but decide to keep looking for a while longer.

Up ahead I catch some green eyeshine, not too high, directly over the road. As we move closer I become more and more excited; a white animal takes shape around the green eyes. I yell for Bill to stop. We take turns holding the light and looking at the possum through binoculars. The fluffy white lemuroid ringtail possum, gently swaying in the breeze, looks inquisitively down at us through the V-shaped fork in a branch. Only his head and haunches are white, or nearly so; the rest is washed with golden yellow. His eyes are dark brown, almost black. His bare nose-tip is pale purplish in colour. It is not an albino, but a golden-white colour variant. We arrive back at camp close to midnight.

At first light a whipbird calls so close that it hurts my ears and startles me awake. A yellow-throated scrubwren is picking up microscopic morsels only a few centimetres from my face. I roll over on my back to take in the new day: no wind, no clouds, a glorious cool morning. A grey goshawk flies low overhead, whistling as it goes.

On our way up to the end of the road to have a look at the high altitude forest, we stop to admire the elegant slender palms we saw in the moonlight. They are even taller than we thought, 15 to 20 metres. This is another endemic. Botanists are still debating whether this is a species in its own right or a Mount Lewis variety of the Alexandra palm, which is widely distributed in northeast Queensland, but usually at very much lower altitudes. The main distinctive features of the Mount Lewis variety are the red-bronze colour of the new leaves and the reddish-purple crown shaft, the area between the top of the trunk and the unfolding fronds. Both are green in the normal Alexandra palm. Around them grow the other, very much smaller, palms. Their luxuriant crowns are equally graceful. The Atherton palm, as it is known, is a high altitude species. Its fruits hang down like long strings of beads. At the moment they are green but in a month or so they will be bright red.

In the balmy sunshine, without the strong wind, the mountain top looks much more benign. The trees, however, do bear the marks of a wind-tossed existence. No emergents project above the canopy; these are soon trimmed back to one even level. Exposed trees have smaller crowns whose branches lean in the direction of the prevailing southeasterlies. Mist as well as wind seems to be the norm here, judging by the growth of lichens, mosses and ferns. Orchids too enjoy the moisture. Many species grow on even the most exposed trees. Buttercup orchids and clumps of *Dendrobium fussiforme* have put out sprays of yellow and white flowers. A still, soft, warm day is the exception here.

The road ends abruptly. STOP has been roughly painted in large red letters on the trunk of a tree. We presume this was addressed to the loggers who were here only a few years ago. the effect of their work is only too apparent up to this point. But a narrow, barely perceptible trail leads into the forest and into a different world. There has been no logging here. The trail leads down a gentle slope to a spring in a valley protected from the wind.

Palms grow in this sheltered valley. As well as the slender Alexandra look-alike and the Atherton palm there is the orania—huge, stout-trunked and stolid by comparison. The other trees too, with their relatively short, but very thick trunks, have this quality of solidness, compared to the slender tall trees of lower altitudes. No doubt the constant wind encourages the trunks to grow in girth rather than height. The undergrowth is more open. Seedling trees are widely spaced.

We cannot identify the large trees, but doubtless there are endemics among them. By looking with binoculars at the crowns of the tall trees we notice that their leaves are small to tiny compared to the generally large leaves of lowland trees. This too is an adaptation to the cold, the mist and especially the constant strong winds.

We remark about the scarcity of vines, especially the hook-bearing varieties. As we stand talking small birds come inquisitively right up to us—grey fantails, brown warblers, mountain thornbills—as though they had never seen people before.

A male golden bowerbird flies across our path. Moments later we hear the bird's ratcheting call, the one he makes only near his bower. We soon find his structure, a mound of sticks piled around four saplings. It is decorated with beard lichen and large white orchid flowers. Finding a bower can be so simple. Around the bower the ground is carpeted with the fallen crimson, tubular flowers of the vine known as misty bells. The flowering part of the vine is so high up that we can see nothing of it.

We have all seen buttresses of many shapes and sizes. Generally speaking they shore up the trees with triangular, bark-covered thin flanges of wood that run down from the trunks to the roots. There is a great variety of buttresses, but what stands before us now is a convoluted, rococo, almost surreal arrangement of thin flexible sheets of wood running from the tree in all directions and for great distances. They belong to a rose alder tree. At their widest part the thin buttresses span a distance of 19 metres. The tree's roots run over ground for many metres beyond that. Most large buttresses impress because of their height, but these do not merely impress, they challenge the imagination with their shapes and the distances they reach out. As they radiate out from the trunk, they curve and twist, fork, and fork again. The flanges also curve in the vertical plane; they descend from the trunk from a height of almost a metre and a half in a series of folds that look like breaking waves. Some have pockets in them. A palm grows out of one of these. The bark of the buttresses is smooth and pale yellowish-green. Despite the fact that there is not a single straight line or right-angle, the whole elegant, lightweight structure resembles a miniature wooden castle complete with ramparts, rooms, courtyards, balconies and hidden chambers; something dreamed up by a most imaginative architect. Bill comments that these are the most wondrous buttresses you could ever wish to see.

Beside them stands a heavyweight of a tree. Its bark is dark red-brown, its buttresses are straight up and down in the classical pattern. Brooding, aloof, 'correct' in its design, this tree provokes a little anthropomorphism. It seems so conservative and disapproving of its fanciful frivolous neighbour.

As is to be expected, Australia's tropical rainforests, which stretch between Townsville and Cooktown, are not uniform; they are not an unending sea of green sameness. Differences in altitude, rainfall, soil, drainage and so on create differences in the forest's structure. The ecologists Geoff Tracey and Len Webb have classified the Australian tropical rainforest into 13 main types. On this scale, number one grows in areas of highest rainfall with the richest soils at lower altitudes where it is always warm. These are places with few, if any, limits to plant growth. The forests they support, not unexpectedly, are the densest, the most complex and, where protected from cyclones, the tallest. The plants have generally large to medium sized leaves. Vines are more numerous and the variety of tree species is the greatest. Bulurru falls in this category. Rainforest type 13, at the other end of the scale, is limited mainly by low rainfall and poor soil drainage. They grow in marginal areas and include certain non-rainforest elements such as eucalypts and wattles.

Where we camped, almost halfway up the mountain, is forest type eight, which is characterised by fewer vines and a generally simpler structure. Higher up the mountain this grades into type nine and on the highest ridges to type ten. These are simpler again with still fewer vines and no strangler figs. Epiphytic mosses, lichens, ferns and orchids are abundant. The higher the altitude the smaller the leaves on the trees.

The Mount Lewis region's isolation from other rainforests means that some species have evolved here. These are the endemic species. Others, like the crayfish, ringtail possums and Alexandra-type palms, have been separated from others of their kind for so long that they have developed distinctive colour phases or other slight differences; they are endemic varieties. Thirdly, there are very ancient and primitive species, like the Mount Lewis stag beetle and the Mount Lewis frog, that managed to survive here and nowhere else. We did not see any of these.

The special character of these higher altitudes which we could sense all around us, the impression of being in a different country, are a reflection of this ancientness, for it is on these granite heights that wet tropical forests have grown uninterrupted since Cretaceous times 140 million years ago. This was a time when Australia was still part of Gondwana, the super-continent that also included Africa, South America, India and Antarctica. Mount Lewis and its associated uplands and the Bellenden Ker–Bartle Frere massif are the only places in tropical Australia blessed in this way. Even when areas now covered in rainforests were stripped of them in the past by long-term droughts or lava flows, rainforest persisted on the granitic heights, where rainfall was always high, and from there recolonised the lands when frequent heavy rain returned and time turned lava into soil.

The tropical wet forests of 140 million years ago were made up mostly of ferns, cycads and conifers. Subsequently they saw the rise of flowering plants. Over time these evolved into an explosion of different species that came to dominate the tropical rainforest and gave it the structure it has today. At one stage in this evolution, perhaps as much as 50 million years ago, most of Australia was covered in tropical rainforest. But only on these few highlands in the northeast of the continent have they persisted without a break. More recently, Mount Lewis and the other uplands absorbed an invasion from Asia of such highly evolved and advanced species as gingers, palms and orchids. Rather than an archaic backwater these uplands are unique in that they contain elements of all the forest types and a significant number of evolutionary steps from the earliest and most primitive to the most advanced and complex, from mosses and cycads to palms and orchids, and from ancient primitive insects to bowerbirds and birds of paradise. As long as there have been wet tropical forests in the world, they have been here.

Living with Insects

PINE CREEK, 23 OCTOBER

Geoff Monteith is an entomologist with the Queensland Museum. Some weeks ago he told me that he was going to the Mount Carbine Tableland to look at the very special insect fauna there. We arranged to meet this afternoon at Pine Creek in lowland rainforest near Gordonvale. Pine Creek is one of the places Geoff will survey for insects.

The creek flows over a sandy bed between large boulders, its tea-coloured water cool and refreshing. Downstream stands a dense grove of Alexandra palms. Some bear large bunches of red fruit. A dozen or so of the white nutmeg pigeons are feeding on them. Above me shining starlings probe the red and orange flowers of black bean trees for nectar, powdering their metallic, iridescent heads with yellow pollen in the process. Green tree or weaver ants are in every tree and bush from ground level to the canopy it seems. Their nests, of growing leaves woven together by silk produced by their larvae, are everywhere. The pale green translucent ants busily scour the forest for prey—usually other insects. An October glory vine, covered in bunches of large white sweet-scented flowers, climbs up and then smothers several trees. About a hundred metres upstream stands a massive fig tree with not just a single trunk, but three enormous pillars rising from a base of buttress roots. The trunks are covered with bunches of large fruit, a spectacular example of cauliflory. The first impression is that all the fruit are red, but looking closer, they vary from deep red through rose-orange, yellow, yellow-green to green. The ground beneath is covered in a layer of rotting, fermenting fruits which are a great attraction to ants. The tree is known as the cluster fig.

Flowers of the October glory vine.

Geoff and two colleagues, Heather and Doug, arrive. Geoff is tallish with a spare

A male Hercules moth. Its caterpillar feeds on the leaves of the bleeding heart tree, so named because of the colour and shape of the old, spent leaves.

and wiry frame. His eyes are kindly and a pale blue. His light brown hair recedes a little from a high forehead and his short beard is a pepper and salt colour. His skin looks as though it has seen a lot of weather; Geoff is no armchair entomologist.

Up on the ridge he has various kinds of insect traps close to a tower that reaches into the canopy. Some are pit traps. These are small plastic dishes set into the soil so that their rims are level with the ground. They are partly filled with a preservative. To stop rain from flushing out the preservative and falling leaves clogging them up, the traps have translucent tops raised a few centimetres above their rims. Insects walking among the litter fall into these traps and die almost instantly in the preservative. Another device is a flight intercept trap. It consists of a sheet of clear plastic set in a vertical frame. Flying insects hit this sheet and fall in a preservative filled trough at the bottom.

The traps have been in place for some weeks and contain thousands of insects, most of them only a few millimetres long. Geoff estimates that this catch will keep him busy sorting for the best part of a week. Another method he sometimes uses is to spread nylon sheets, about two by three metres in size, around a tree or at the base of a large rock and then spray the area with short-lived pyrethrum insecticide. All these methods yield insects in variety and numbers not possible with the classical methods of sweeping the butterfly net or picking insects off by hand, though those techniques are also used.

The tower is a daunting structure, a series of ladders leading to ever-higher platforms anchored by enormously long steel cables. I have no head for heights but I am determined to make it to the top.

Moving slowly from ground level to the canopy is a wondrous journey. First we pass the crowns of the fan palms; some of the fan-shaped leaves, the size of cartwheels, brush up against the tower. Looking up from the ground, you are aware of the palm's trunk; it looms large and solid even if somewhat slender. From here you are aware of the huge circular leaves on long stems forming a rosette, while the trunk looks far too thin to support such a mass.

Next we reach the thick lower branches of the canopy trees. We are face to face with, then look down into, large epiphytic ferns, the staghorn, elkhorn, bird's nest, and basket ferns. We can see the leaf litter that has accumulated in them and has formed into actual soil. Geoff suggests an especially large bird's nest fern might have been a good tree-kangaroo's den had we been on the Tableland. (There are no tree-kangaroos in these lowlands.) This conjures up mental pictures of these marsupials leaping along the branches at the level at which we are now standing. To me that seems a very unstable world; everything sways and rocks in the wind, bends under one's weight, branches break. But I also envy the tree-kangaroos their ease and nonchalance at these heights, and this magical perspective of the forest.

For once I am not picking fruits up from the ground and then straining my neck and eyes in an effort to see what tree they came from. I can reach out and pick the large yellow-green fruits right out of the topmost branches of a species of walnut. A little higher and on the opposite side of the tower bunches of orange fruit of a native species of tamarind hang within arm's length. We are now in the lower parts of the crowns of the tallest trees and among the huge clumps of white bottlebrush flowers and shiny dark green leaves of a northern silky oak. But even here we are not out of the grasp of the wait-a-whiles. They claw their way up from below and continue on above to the very tops of the trees. Looking down on these climbing palms, seeing their feathery leaves arranged in whorls around their prickly stems, gives them an added and unexpected symmetry and beauty. A pepper vine also snakes up and spreads its foliage and bunches of red berries around the tower. Given time, the tower will no doubt be the support of several climbing plants.

At last I can appreciate what it is like to be in the rainforest canopy. On three sides we are level with the highest branches of the tallest trees. On the fourth side, which faces the east, we look down over treetops, across a valley to a forested hill. There are many places where you can look across a rainforest valley, but few where you can do so while standing in the tree tops.

What is most noticeable up here is that every tree crown has its own identity. Some are flattish, others rounded or dome-shaped. Each has leaves of a different size, texture, shape and shade of green. When the breeze picks up, each crown moves independently. One tree may be motionless while another sways in a wide arc, depending on how many of the leaves catch the breeze and how flexible its trunk is. It is also quite apparent that the trees do not truly interlock. Each moves within its own space without intruding on its neighbours.

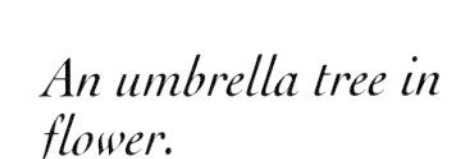

An umbrella tree in flower.

Looking down to the forest floor, the young Alexandra palms are like delicate feathery circles. The fan palms appear coarse by comparison. The leaf litter is a rich warm brown in colour, a harmonious complement to the many shades of green. Looking down is much more restful to the eyes than squinting up into a bright sky. From down there the tree trunks look solid, many of them rigidly buttressed. Up here they are slender and rock gracefully back and forth.

It is late afternoon. We are still in the sun but the ground is in shade. Suspended, as it were, in the treetops, floating on the semi-dark below, seeing the trees sway gently all around, I suddenly have a feeling of spaciousness, of being in a truly three-dimensional world. It is a sensation I have never experienced in the forest before.

Up in the canopy the forest also makes more sense; it appears more vital and energetic. It is driven home that up here the real business of the forest is conducted. Reluctantly we descend. With every step down the magic is lowered a notch until we are back in the normal, everyday world.

It is nearly dark when we get back to camp, but the insect collectors' work is not yet done. Geoff sets up a small generator at the edge of the forest to run a mercury vapour lamp. The lamp's light, which is high in ultra-violet rays, is irresistible to a great many night-flying insects. Moths, beetles, bugs, katydids and others land in their hundreds on a sheet stretched across a frame placed behind the light. While giving insects time to come to the light, we retire to camp. We use just a small light run off the car battery. Something large with slow-flapping wings flies in among us and circles the lamp. My first thought is that this is a tube-nosed bat. But it does not fool the entomologist. Geoff knows immediately that a female Hercules moth has flown into camp. These moths may have a wingspan of up to 25 centimetres and are the largest in the world. While this enormous insect flies around us Geoff points out how it keeps its body in a vertical position, not horizontal as most insects do. She is an old and tattered individual, close to the end of her life. She must have been laying eggs; one still adheres to her abdomen.

Moonlight floods the forest when we set off to find what insects can be gleaned from small trees and shrubs. But first Geoff checks the ultra-violet light. The sheet is speckled with small to tiny insects, a great many of them moths that resemble flakes of bark, small sticks, pieces of lichen or moss. On the plain sheet some are jewel-like in their bright greens and greys or browns and yellows, but on the trunks of trees or the dead leaves on the ground they would be impossible to find. A few of the small moths disdain camouflage; they are bright pink, yellow, or white with red trimmings. A great many other insects sit transfixed by the light, but Geoff judges it to be a poor night. At this driest time of year there are probably the fewest number of insects about. In six weeks or so, when the first thunderstorms of the season drench the forest, there will be so many flying insects that they would completely cover the sheet.

Geoff, Heather and Doug strap lights to their heads to leave their hands free for gathering insects. I carry a torch. We cross the creek. A giant frog, so large it would easily cover the palm of my hand, clings sideways to a sapling and blinks its orange eyes. With their headlights the insect hunters examine bushes, grasses and tree trunks minutely. They pick off species of interest with forceps and put them into bottles of preservative. A few spiders are also collected. We pass a rufous shrike-thrush soundly asleep in a shrub beside the trail. Geoff sees a snake's tail slowly disappear into a mass of vines. With his light he traces the reptile's coils, only to discover that the amethystine python's head is just a few centimetres from his ankles.

The moon, yellow and full, appears from behind some clouds when we reach the tower again. We decide we should see the canopy by moonlight. There is a great feeling of peace up here. Not a breath of wind so much as rustles a leaf. A tuft of cloud hovers over the hill top. Pools of moonlight illuminate some fan palms below. Only the sounds of crickets interrupt the silence. But even at night the weaver ants scurry about.

Back at camp we build a small fire and brew some tea. Geoff and I talk deep into the night about rainforest, insects and Gondwana. Geoff has been a rainforest entomologist all his adult life. He grew up on a dairy farm near Wondai in southern Queensland where, he says with a certain irony, he was taught to shoot and kill things. Although he comes from a family of six, he always thought of himself as a rather solitary kid who was happiest when he was in the bush. It was an interest that stayed with him and shaped the course of his life. Strangely enough, his path to the bush began in the city.

'I always wanted to do something in the bush. When I was fourteen we moved to Brisbane and the possibility of going on to further education loomed. The only inspiration I had then was to go into forestry because I thought foresters must work in the bush. At that stage I didn't know that zoology existed as a profession. So when I got to the university, intent on doing forestry, I discovered there were whole departments of zoology. So I switched. What turned me to entomology was a couple of months in New Guinea where, as you know, there is a lot of rainforest. At the end of my second year at university I worked there for the Bishop Museum at their field station at Wau. I discovered insects mostly under the influence of a Czech couple, Joseph and Marie Sedlacek, who ran the place. They certainly had a passion for insects, and still do. I think I have more of a passion for being in the bush, for being away from cities. Insects are very much a means for getting into the bush. Well, that's not quite true. I *do* have a passion for insects. I don't do anything else but think about them. And rainforest is where I've always headed for. I'm not sure why. Perhaps it began because the particular insects I'm interested in, the ones for which I work out the formal taxonomy, are the fungus bugs which are almost completely rainforest things.

'I very clearly remember my first experience in the tropical rainforest up here. I came up on my own. I would have been twenty-one or twenty-two. I had just completed my third year at university and got my first car. I hopped in it and headed for north Queensland for a couple of months over the Christmas break. The first place where I spent any time was Paluma. Right from that first time I followed on the trail of the American entomologist Philip Jackson Darlington Junior. He had spent all his life trooping around rainforest. First in the West Indies and then during World War II in New Guinea. He made enormous collections in New Guinea and he wanted to come to Australia to see if there were any similarities in the insect faunas. He travelled from Tasmania to the tip of Cape York, but spent most of his time here, in the wet tropics. He was the first to climb all the major peaks in this area and collect insects there. He climbed Mount

Finnigan, Thornton Peak, Bellenden Ker, Bartle Frere and others in the late 1950s, well before they had beaten tracks up them. He discovered a whole unknown insect fauna of quite large species. He was especially interested in the large wingless carabid beetles.

'Darlington has been a great influence, not only because he collected and described all these interesting beetles, but also because he published a paper which was a list of places he went to during his eighteen-month trip to Australia. It gave details of all the localities he went to, what sort of rainforest was there and what insects he found. It was sort of my bible; that and the account of the 1948 Archbold Expedition. That narrative, by Len Brass, is one of the great unrecognised classics of descriptive expedition work in Australia. That expedition too had been up most of the major peaks, although insects were not their primary interest.

'So I followed in their footsteps. I went mostly from one camping place to another. I used to sleep on the picnic tables whenever I could. On that first trip I did a great deal of light trapping. Light trapping in December is phenomenal. That's when the early storms are around, when there are massive emergences of things, when all these thousands of colourful beetles come out. They were spellbinding nights; just by myself running my light and my generator. I had a great big heavy generator in those days that filled half the boot of my car.

'Backpacking into uncharted rainforest for nine or ten days has been a real awakening for me in the north here. To walk into places like the Mount Carbine Tableland or Mossman Bluff is a stunning experience. Once you get out of the wait-a-whiles of the lower altitudes you're in benign mountain rainforest.'

Geoff and his companions have just returned from two backpacking trips. One was to the Massey Range near Mount Bellenden Ker, and the other to the Mount Carbine Tableland; still in the footsteps of Darlington and the Archbold Expedition. I ask if they had found anything of special interest.

'One of the really nice things,' Geoff says, 'was that we found a female of one of the strange fireflies [which are really beetles] that have wingless females. Because they don't fly, females are rarely seen and are even rarer in collections. To find them you have to look for a flashing light on the ground. It's her signal to the males that fly around. As I reached out to catch the first female, I saw a male plummet out of the sky and land about five centimetres away from her and then make his way manfully towards her. So the signal got through. The female, which is a couple of centimetres long, looked just like a queen termite with a great bulging white abdomen which contained the flashing light. That was one nice thing. The other involved king crickets on the Massey Range.

'We found a pair of king crickets mating. After the first part of the mating, after the male has put his spermatophore, a kind of package of sperm, into the female, the pair stayed in copulation for many, many minutes, during which time the male pumped out a pea-sized droplet of a white gelatinous looking material. This stuck to the female's abdomen after the two finally separated. She then reached down and ate what was obviously a nuptial gift from the male. To ensure his genes will be carried on, he gave the female a big feed so that she'll have the strength to nurture his young. From memory this behaviour hasn't been noted in king crickets at all yet. This particular species belongs to an undescribed genus.'

Inevitably, during his many backpacking trips Geoff must have experienced periods of rain. I ask how he copes with that.

'Oh, we keep working. I've always had the attitude that when it rains plans go ahead regardless. With my workplace a couple of thousand kilometres away, it takes a great deal of expense and effort to get here. So if it happens to rain on the day we plan to do something we must go ahead. We've been through some horrendous weather. I can say, though, that I've only had my camp thoroughly wet once. That was on Mount Elliot near Townsville. We set up camp in a dry creek bed. Suddenly it began to rain. It wasn't the wet season at all, yet it rained and rained for three days. The creek rose and we kept moving camp further and further uphill. We ended up with totally wet gear on that occasion. But otherwise we've survived. We always put a tent fly up and we've never yet failed to be able to have a fire to sit around, no matter how heavy the rain has been.'

I ask Geoff why he has concentrated his insect searches on the highlands.

'It is almost like going to two different continents when it comes to lowland and highland insects. The overriding factor in the making of the two different insect worlds is the ants. There are ants on the highlands, of course, but there aren't very many different sorts. In the lowlands they're totally dominant. The result is that there are whole suites of species of insects that just don't exist in the lowlands because they cannot compete with ants. That seems to be the story with the carabid beetles, especially the wingless species that Darlington was so interested in. They're all confined to the highlands, and each mountain evolved its own species because they couldn't fly from one peak to another, and if they walked they couldn't get through the ants. Also ants are mostly predatory and so are the carabids. The beetles just can't compete with the ants and are therefore absent from the lowlands. There are quite a number of other insect groups like that and they happen to be the ones I'm most interested in. The mountains also harbour most of the primitive insects with Gondwana links and that too is of great interest.'

That makes me prick up my ears for it is yet further evidence of the great age of these forests. Insects too have a long evolutionary history, going back even further than that of the flowering plants. Like the plants, primitive species among them have close allies in places that were joined together when Gondwana was a single supercontinent. I ask Geoff to tell me something about the Gondwana insects. He replies: 'The Mount Lewis stag beetle has a classic Gondwana distribution. I've personally never collected one, but we have four at the museum and it has now also been taken on the Mount Windsor Tableland. The first one was found on Mount Lewis about ten years ago. To our astonishment it belonged to a genus of about twenty species until then known only from the Andes mountains of South America. Then there is a strange flightless leafhopper that belongs to a genus that occurs here and in Madagascar, New Zealand and Chile.

'Let me tell you about one of the most extraordinary field outings I've ever been on, when I turned up some other Gondwana things. It was just a trial trip to Cape Tribulation. I was staying at Pilgrim Sands and had my family with me. One day I went off by myself and scooted up the main ridge behind the Cape to the top of Mount Sorrow. Among other things I brought back a bag of leaf litter. I had my extraction gear with me [for extracting minute insects from leaf litter], as I always do on holidays, and a microscope. Out of that one sample of stuff I got terrestrial water beetles, terrestrial waterbugs and the terrestrial nymph of a dragonfly. These all belong to groups that usually live in pools and ponds, that is in water. But those were all in leaf litter on solid ground. The dragonfly nymph is very strange. It looks like an ordinary nymph that you've stood on and squashed flat. This flattened shape seems to be a modification for adhering to the moist surfaces of dead leaves. It's also very well camouflaged and the sort of thing you would never, ever find just by looking for it. It is quite a rare dragonfly and the nymph had not been found before. These insects can survive in the litter because it is so wet up there. [The nymphs, or immature stages, of all other dragonflies live in water.] If records were kept

The flightless carabid beetle Pamborus opacus *has curved, toothed mandibles to grip such slippery prey as earthworms and snails. Carabids belong to a whole suite of primitive insects that live on the high mountains of the wet tropics.*

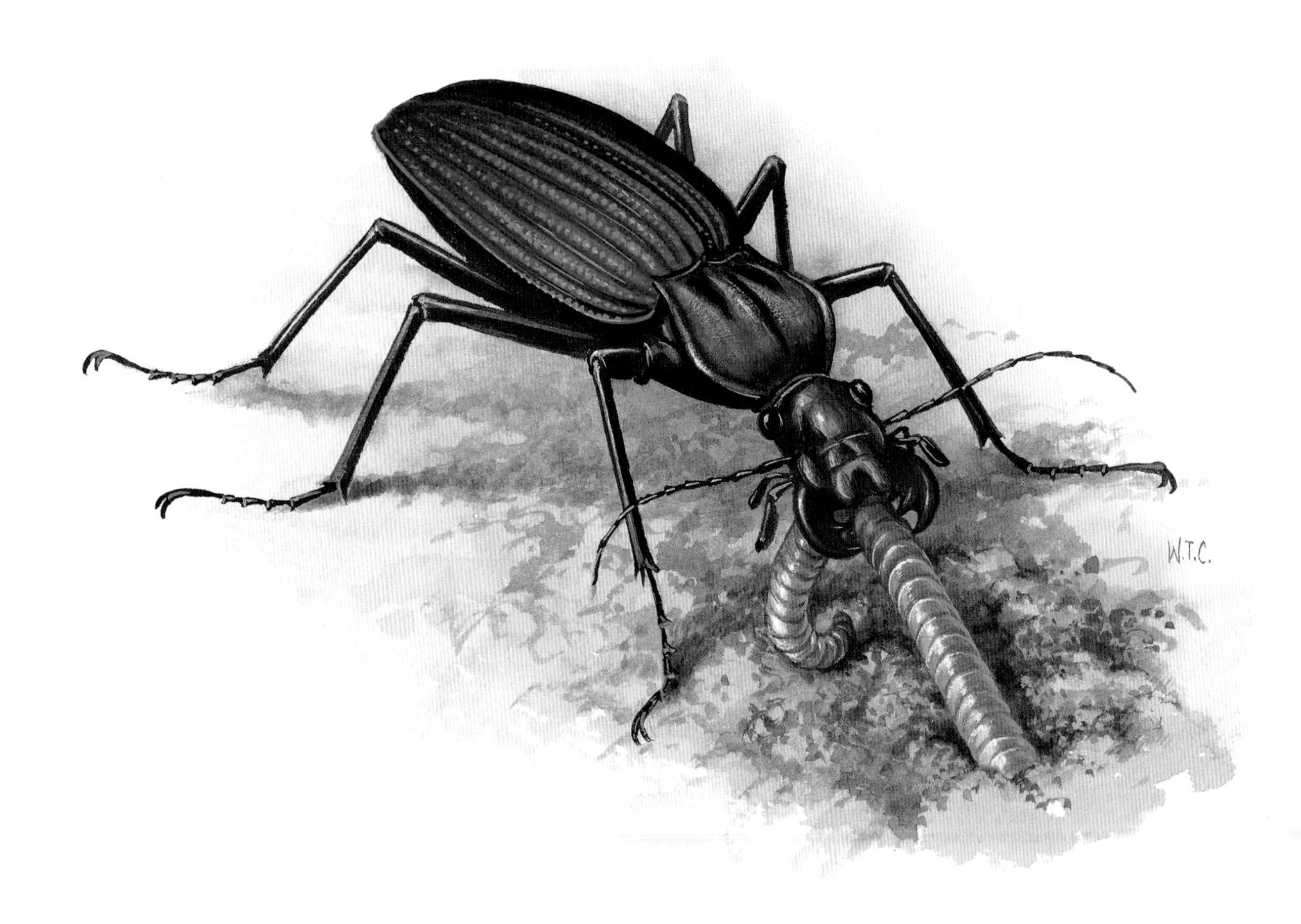

on the hills behind Cape Tribulation you'd probably find that it's wetter there than on the top of Bellenden Ker, where there is a rain gauge. Bellenden Ker has an annual average rainfall of 9000 millimetres, I think, the highest in Queensland.

'The insects I collected that day also have a Gondwana distribution. The terrestrial water beetles are found only in the Cape Tribulation region in the wet tropics, nowhere else in Australia, and then the highest mountain of New Caledonia and in the Himalayas of Nepal. The same for the terrestrial water bug. The dragonfly is the only one in the world with a terrestrial nymph.'

We talk a little about how the Gondwana connections were first discovered by forest ecologists working on plants. I ask Geoff if he is interested in plants.

'Yes, more and more,' he says. 'I regret my university training very much in that respect. It taught me heaps and heaps of botany but nothing at all about plants. I learned many things about sections of stems, cell structures and flower dissections, but never anything whatsoever about the native Australian plants. It's only in recent years that I've spent a lot of time working on plants. Any entomologist must know plants or he never knows what his insects are doing.'

Geoff is fascinated by the ancient and profound Australianness of our tropical rainforest, something that has only been unravelled in the last 20 or 30 years. He admits that his university studies left him unprepared for that as well.

'Like so many biologists who went to university in the 1960s, I was told that our Queensland rainforest came from Asia, trickling down via New Guinea. So all the animals and plants here should be related to Asian and New Guinea groups. That is true, to a degree, for the lowlands especially, but as soon as you get a little elevation under your feet, the story changes.'

That history, I interject, was rewritten by forest ecologists like Len Webb and Geoff Tracey. They changed our thinking about these forests and their origins. Geoff nods in agreement and adds that the insects are, of course, so very closely linked to the plants. I ask what it felt like to come to such revolutionary conclusions.

'Well, I have to admit it didn't come in a flash of light. It's been a gradual thing, for the botanists too. It was Len Webb who told us, and he had many philosophical reasons for saying these sorts of things to Australians, that we shouldn't be thinking of our rainforests as some migrant thing from Asian lands. We should be thinking of our rainforest as something that evolved here in Australia, as a truly Australian ecosystem and as a very ancient ecosystem. All of which has now been verified. We had been told for years that eucalypts, the gum trees, form the archetypical Australian habitat. Well, they don't. That's nonsense. Eucalypt habitats are very recent compared to rainforest. They may be widespread but they're very new compared to the things we see when we climb to the top of a mountain in the wet tropics.'

Teeth Almighty

Bulurru, 3 November

With its expressive ears, intelligent face and long black whiskers, the white-tailed rat is a most appealing animal. Even large males, which tend to be battle scarred and noisily aggressive, have a certain roguish charm. It is their teeth that get them into trouble; nothing is safe from this large rodent's gnawing incisors.

Last night I left my gumboots outside the front door. This morning there are puncture holes in the toes and the telltale toothmarks of a white-tailed rat all over them. For the last year or so I have been free of the attentions of these rats. I will have to catch this raider alive and, like the other half dozen or so I have caught, release it in the forest a long, long way away.

White-tailed rats, or giant tree rats as they are sometimes called, were the bane of the early settlers, timber getters and prospectors living in or near the rainforest. The rats could ravage a camp overnight, stealing food and gnawing through everything—leather, canvas, rope, rubber, wire netting, tin plate and latterly all kinds of plastic. Any plastic waterpipe left above ground is soon perforated. Machinery, especially that with hoses and

White-tailed rats often eat the fruit of Atherton oaks.

belts, is sabotaged. The powerful incisor teeth are efficient can openers and also rip through the lids of glass bottles. No matter where you may hide food, the rats will get to it, whether it is in the rafters or under the floor-boards. When the Coopers were camped at Chowchilla before their house was built, they were woken one night by one of these rats. They were just in time to see it climb up a tree, their Vegemite jar clamped under its chin, and then disappear into a hollow.

White-tailed rats perfect their gnawing skills in the forest where they seem to specialize in extracting the kernels from the hardest seeds, some of which are truly as hard as rocks. When out in the forest at night I have heard and seen them gnawing native walnuts and the thick shells of Whelan's silky oak. Hardest of all could well be the nuts of the ebony heart tree, a kind of quandong. The first time I saw these beautifully shaped and textured nuts lying on the ground I thought they were fossils, they were so heavy and hard. They also had the colour and texture of stone. But each one I picked up had a hole gnawed through it and the kernel extracted by one of these large rats. The kernels taste good and are nutritious. They were much prized by the Aboriginal people.

The Atherton oak, a favourite of the rats, is in fruit at the moment. After dark I go out with a spotlight to see what the rats may be up to. Long before I get to the tree I can hear one gnawing. The sound comes from high up in the tree. Bunches of the shining blue fruit catch the light. It takes a few minutes to locate the rat. He is sitting on a large horizontal limb, a fruit grasped in his hands. Within a few seconds he strips the flesh away, a few more gnawing bites and he is through to the tasty centre. the Atherton oak's nuts are not as hard as those of the Whelan's silky oak or ebony heart. I hunt around the bottom of the tree for some nuts that might have escaped the rats' attentions. I can only find a few which by the look of them have only just fallen or been knocked down. I put them in my pocket for I am also quite fond of these kernels.

But unless I climb the tree I can never get many. Hundreds upon hundreds of the seeds of several seasons lie scattered under the tree and I would wager that apart from the ones that I intercepted tonight, not a single one is whole. In the past, out of curiosity, I have sifted through the fallen seed crops of several large Atherton oaks. Among the thousands I examined only two or three were still in one piece. Yet enough seeds escape these teeth almighty for the trees to regenerate. If not the Atherton oak would have been extinct long ago.

Fruit of the Atherton oak.

Vic Stockwell's Puzzle

Boonjie, 24 November

Two weeks ago the first storm finally arrived. There have been several more since and yesterday afternoon a spectacular one hit Bulurru. Huge white clouds piled up in the south at about midday, grew larger and larger and turned a dark indigo. When the thunderstorm was directly overhead it broke. Lightning speared down on the next ridge and sparked from cloud to cloud. Deafening thunder rattled the windows. Rain bucketed down and was driven almost horizontally by fierce winds. Branches were ripped off trees and flew through the air. A bollywood near the house was blown down. The gutters overflowed in sheets of water. I felt like I was behind a waterfall. It was all over in an hour.

This morning is cool and fresh in the Boonjie forest at the base of Mount Bartle Frere, only a few kilometres from home. I walk along a sometimes ill-defined trail that follows an old logging track. In places the wheel ruts are still visible even though logging ceased 15 or more years ago. There are no truly large trees left and I pass the stumps and discarded logs of the more durable species. Ones with softer wood have long since rotted away. It is an all too familiar pattern in so much of the rainforest.

But after half an hour's walk I climb up a steep pinch and come to a small plateau where everything suddenly changes. I am at the edge of a completely different kind of forest, one dominated by trees of enormous size and all of them belonging to the same species. Other trees grow beneath the giants' wide crowns, and while some of these are sizeable they all look spindly by comparison. It is the clearest example of a forest within a forest. The undergrowth is a rather open combination of seedling trees, ferns, the occasional wait-a-while and walking stick palms. Down the gen-

Lichen and moss after rain.

Most splendid of Australia's innumerable beetles is Mueller's stagbeetle, found only in tropical rainforest. Its larvae feed on the decaying wood of fallen trees.

tle slope there is column after column of the large trees. To my right three grow closely together, to the left, some way down the slope, two stand side by side and dotted throughout, 30 or 40 metres apart, grow single trees. There has been no logging here. I had been told about these trees, but I am unprepared for the sheer spectacle. Life is on a different scale here. White-throated treecreepers, so prominent in the trees beside my verandah, here look like tiny pinpricks going up the big rough-barked trunks. A rufous fantail dancing among some shrubs, its warm red-brown colours a paler shade of that of the trees' bark, looks lost and insignificant between wall-like buttresses. I can just make out the tiny yellow splash of a golden whistler in the canopy. Smaller birds, mountain thornbills perhaps, look like insects flitting among the leaves. On the other hand, the clear notes of the whistlers and shrike-thrushes carry further and have a greater resonance.

The first tree I go close to is a scarred, ancient-looking veteran. Its straight trunk is of enormous girth, but ends abruptly only about 10 metres up. It had been snapped off many seasons ago in some storm or cyclone. Since then it has coppiced, sending up four or five smaller stems. In another assault during more recent high winds several limbs have been broken off another tree. These lie scattered in a tangle around its base. The bark has rotted away. The exposed wood has a tight wavy grain, very like that of certain eucalypts. A dozen metres away stands a tree without any blemish whatsoever; it is a straight pillar going up and up to branches so large that a man could lie comfortably across them. Each tree is different, each has its own special character attained over many centuries. One solid trunk has a lean on it and its buttresses are covered in knobs and bumps. Its neighbour's trunk is fluted and has elaborate buttresses that grope six metres or more downhill. One buttress seems to be tied in a knot. Twenty metres away stands an old ruin of a tree, broken and gnarled on one side. Several buttresses have died and collapsed into rubble. Yellow fungi grow out of them. But the other half of the tree is still alive, continues to grow, and has reached the canopy again. Not all the trees are gnarled and ancient. Young and vigorous ones grow in small groves among their elders.

Each tree I go to seems to have taller, longer buttresses than the one before. Eventually I come to the one that must have the ultimate. The wooden walls rise out of the shrubs to a height of four or five metres. Curving away from the tree, they form a forecourt to the trunk itself, which is hollow. There is a natural doorway into it and I enter without having to stoop. I look up this hollow chimney and see the tree's foliage against the sky. A bat circles above. There is a back door as well. Stepping through it, I come to another courtyard with high wooden walls.

This tree, so large, so ancient that it is difficult to think of it as a single structure risen from a single tiny seed, is less rococo than the wooden castle at Mount Lewis. It is also far more robust and on a grander scale; it is more a massive fort than an airy castle. Beyond this fort a narrow ridge runs to the south. The carcass of an ancient tree lies across it. Its bulk has almost rotted away. Only a few sheets of moss-covered wood remain above the soil. All over and around it thousands of tiny seedlings, of the same species and only a few centimetres tall, have come up, their roots feeding on the decaying colossus.

In all the places I have been I have seen no greater trees, nor such an overwhelmingly grand forest. These trees are today not yet named or described by science. They have been found only on these few hectares at the base of Mount Bartle Frere. Although these enormous trees may occur in other places, it is unlikely; they are not the kind that go unnoticed. They are almost certainly found here and nowhere else on earth. They came to be noticed in the late 1970s, during logging operations. The Forestry Department's inspector for the area, Mr Vic Stockwell, discovered they were not on the list of species that could be cut. He knew straight away that they were new to science. As a result he would not allow them to be logged. He showed the trees to botanists and forest ecologists, who scratched their heads. Where do these trees fit into the scheme of things? It was a puzzle. Such an enigma that the trees were given the interim name of Vic Stockwell's puzzle. It was recognised, however, that these trees had to be saved and logging was stopped in the area. While still not described, the trees' special characteristics have been recognised.

There is something vaguely familiar about them. The colour and texture of the bark, the grain of the wood, the shape and size of the glossy leaves, the small seed capsules, are more like those of a forest eucalypt than a typical broadleafed rainforest tree. But they looked most like a tree in the monsoon forests of the Arnhem Land escarpment in Kakadu National Park in the Northern Territory. That tree was named *Allosyncarpia ternata* by botanists in 1976. Stockwell's puzzle and *Allosyncarpia* are indeed closely allied species belonging to the myrtle family, the same family that includes the eucalypts, paperbarks, bottlebrushes and many other species that dominate Australia's dry country habitats. It is now thought that these two species of giant tree, one from Kakadu, the other from near Mount Bartle Frere, and perhaps a few other related species, are the progenitors of all the eucalypts and other Australian members of the myrtle family. We come back to the same phenomenon that was so strongly evident on Mount Lewis: tropical rainforest is the most Australian of this continent's habitats. Rainforest is not a recent invasion from Asia. What are now considered typical Australian forests and woodlands, dominated by eucalypts, banksias and other species of the myrtle and silky oak families, the species that give an Australian feel and character to so much of our landscape, are derived from these very rainforests. Their precursors come from here, their origins are in ancient tropical rainforest, something that now remains only in northeast Queensland.

Nutmeg Pigeons

Low Woody Island, 3–4 December

The nutmeg or Torres Strait pigeons returned from Papua New Guinea and other islands to the north during the first half of August. We see the odd one on the Tableland, but it is really a bird of the lowlands. They gather in large numbers wherever there are fruiting trees. I well remember a day in October, near Cape Tribulation, when nutmeg trees and a kind of laurel were coming into fruit. The breeding season was well advanced by then and several thousand of the pure white birds were courting in a grove of trees. The males would puff themselves out till it seemed they must burst and called 'roo-ca-hooo'. The final 'hoos' boomed and reverberated through the forest. While they court on the mainland, the pigeons do not nest there. Every afternoon all the breeding birds fly to one of the offshore islands and their nests, leaving the forests suddenly quiet. The pigeons nest in colonies sometimes numbering many thousands of pairs. It is to see this spectacle that I travel to Low Woody Island, about 15 kilometres from the beaches of Port Douglas.

Early this afternoon a friend picked me up in his boat. Nearing the island, he cuts the engine and we glide towards the shore. I can now hear the pigeons in the mangrove trees, an occasional booming 'roo-ca-hooo' rising above the general murmur of gentler oos and ooms. Low Woody is 21 hectares of coral rubble held together by a dense mangrove forest. Not much rises above high tide mark. When I start slapping at sandflies my friend laughs, wishes me luck and is off. He promises to return late tomorrow. The sound of the boat's engine soon fades to nothing. I am alone with the pigeons and other birds that make their home here.

I walk to the northernmost spit where there are no trees. It is low tide and I sit for a while on a pile of driftwood. Even though the sun beats down fiercely, I am cooled by a soft breeze. I can see the pigeons' entire domain. To the north and west are the rainforest clad mountain slopes and coastal flats stretching from Mossman Gorge to Cape Tribulation. These are the places where the birds feed. At the other end of the spit, just a hundred metres away, starts the mangrove forest where they conceal their nests. In between is a stretch of water, varying in width from 12 to 20 kilometres, that they must cross every day. The blue mountain ranges are topped with the towering white clouds of a gathering storm.

A great deal of care is always needed when entering the nesting colony of any species. Birds rushing off in fright may dislodge eggs or small young. When birds are kept off their nests, these are easy targets for a whole host of robbers from eagles to goannas. I enter the mangrove forest gingerly. I follow the island's eastern edge where there is a narrow corridor of higher ground. Walking is easier here. Some of the mangrove trees have smooth bark, others are rough barked; some have large leaves, others small ones. There are mangroves on stilt roots and species without them. The pigeons nest at all levels of the forest except the very tops of the trees. With every few steps I take I can hear the slap and clatter of birds flying off their nests. In all directions I glimpse their white forms through the dense foliage, so dense that I can see only about four metres ahead. Luckily I can hear, and see, the birds return to their nests behind me. If I had not, I would have had to abandon my walk.

Nests, not all that closely spaced with three or four per large tree, cover the entire island. They vary in size and method of construction. Some are no more than a handful of dead sticks put seemingly haphazardly together in the manner of most fruit pigeons. Others have solid bases made of leafy twigs. These the birds must pick green for some are still fresh. Some nests are just being built while others contain a large lustrous white egg. Young, from helpless pink little things to those almost ready to fly, occupy other nests. One nest, built on the topmost stilt root of a mangrove, contains a well feathered young, white like its parents. It regards me calmly through black eyes. I reach out to stroke it. When my fingers just about touch its back, the young bird hits me, hard, with its wing. In the high branches of one mangrove five complete nests have been built one on top of the other. A large young sits on the topmost.

On the rims of the older and larger nests, and beneath those where the ground is not constantly washed by the tides, are accumulations of seeds, voided by the pigeons. I recognise quandong and nutmeg seeds. Many have germinated in protected corners, creating small green carpets. But there is no future for these young plants. As soon as their roots reach the salt water, they will die.

There are few empty nests, suggesting that there is little interference

Nutmeg pigeons feed in lowland rainforest but nest on off-shore islands.

from predators and that breeding success is high. There are no signs of snakes, goannas, butcherbirds, rats or other nest robbers. Nesting on islands has its advantages and may well outweigh the effort of flying 20 kilometres each way to the mainland to feed. As long as there is a ready supply of fruit it works well. Between September and the end of January some pairs will have raised three young in succession. Many more will have raised two and the majority a single young. Storms and cyclones that can uproot trees and destroy nests seem to be their main enemies.

Eventually I emerge from the mangroves at the other end of the island. As I push through the foliage of the last trees I look up to see a white-breasted sea-eagle watching me intently from his perch in a dead mangrove. There is a predator on the island after all. Healthy vigorous adults have nothing to fear from the raptor; they can easily elude him. But he probably does take large young from some nests. On the edge of the tall forest grows a row of shrubs with small leaves. These too are mangroves. On the ground beneath them scores of freshly broken twigs lie scattered.

This is where the pigeons come for their nesting material.

Not wishing to disturb the pigeons again, I walk back along the tideline. Near my camp, such as it is, I gather some driftwood, including a few well-preserved pine planks, and construct a rough hide. The driftwood provides the frame which I cover with cloth I have brought with me. From this hiding place I hope to observe what happens when the parents return to their nests.

By the time I resume my seat on the northerly spit, the white clouds over the mountains have grown larger and darker. A stiff breeze blows in from the southeast. It is about two and a half hours before sunset and I suddenly realise that the arrival of just the odd pigeon now and again has been transformed into a steady trickle. They arrive in threes, fours, sometimes a dozen. Birds coming in must beat against the wind and fly just centimetres above the dark blue choppy seas. Flying fast and straight they easily overtake crested terns coming to nests on another part of the island. When the pigeons see me they sweep upwards, passing high overhead. Most land in the tops of the mangroves where they sit a while before going to their nests. Perhaps if I hide behind some driftwood, which is piled high in places, the pigeons will fly close over me. It works. Flock after flock comes in and skims just a few metres over my head. I can hear the whoosh of their wings above the sound of the surf. Their crops bulge with marble-sized fruits and the sun glints in their eyes. I notice that their flight feathers are black.

The volume of birds steadily increases. Many flocks number more than fifty. One flock, approaching in a wide V formation, consists of 163 pigeons. Birds coming from a long way away are like white dots against the dark sea and darker mountains. The few birds leaving the island fly high and swift, the wind in their backs. The volume of ooms and coos swells and is now reinforced by the booming of amorous males. The whole island throbs with sound. The sun slips behind the storm clouds. Few scenes in north Queensland are more elemental and dramatic: a rough sea, a gathering storm, the rush of thousands of powerful wings, the sound and energy of birds courting and raising their young.

During a 20-minute period I count 2200 birds arriving. That means that over the two hours of the main influx 13,200 nutmeg pigeons fly in. Assuming that the arrivals have mates looking after nests, close to 26,000 pigeons are on the island at nightfall. And that is not counting the young.

Towards dusk I leave my vantage point and sneak into the hide. Pigeons on nests all around take little notice. In the early afternoon there were only quiet sounds. Birds on their nests yawned, stretched their wings, nibbled at sticks, snapped at sandflies and hummed an occasional 'oom'. Quiet boredom. Now the colony is galvanised. There is a constant clatter of wings of birds landing in the tree tops. Calls are passionate. I can also hear strange nasal sounds, something like 'cleck'; the call of one bird greeting another at the nest. Everywhere snow white birds move through the dark green forest. One lands heavily on top of the hide. I lean forward and put my eye to a tear in the cloth and look straight into a pigeon's gaze. I take it to be a female because of her slimmer build. She can probably not see me in the darkness of the hide. She flies to the nest tree in front of me, hops from branch to branch, walks nimbly along a thick limb and with head held high walks onto the nest right in front of me. An actor of great dignity could not have made a grander entrance. Gently she pecks her young who moves under the protection and warmth of her breast. Unlike the nestlings of most birds, it sits quietly without noisily clamouring to be fed. She will soon feed it. But darkness is near and lightning dances in the west. The rising tide laps around my ankles. It is time to set up camp.

I manage to find a few square metres of space out of the reach of high tide. I do not sleep much. A storm sweeps in and drenches just about all I have with me. Luckily it is not much. When I hear the wind ripping through the mangroves I realise why there are no nests high up in the trees.

Daybreak is calm and clear. Judging by the quiet sounds from the nutmeg pigeons the mood there is one of contentment. One by one they gather in the tops of the tallest mangroves till the trees are white. No booming calls this morning. Birds doze in the sunshine. Some sunbathe; leaning over and lifting a wing, they let the warm early sun shine on its underside. After a few minutes they turn around and expose the underside of the other wing. I hang my sleeping bag and most of my clothes out to dry.

The sun is well above the horizon when a few birds lift off and, rising high, speed towards the mainland. As one lifts off from one tree others join it from nearby vantage points. Flocks of thirty, forty, sixty race off. Soon the tree tops are empty of nutmeg pigeons.

Fan palm. Fan palm swamp forest is one of the many types of tropical rainforest protected under World Heritage listing.

World Heritage

CAIRNS, 9 DECEMBER

All this magnificence that is rainforest, what is its future? Nowadays when we are captivated by a natural area, when we recognise it as a vital part in supporting life on this planet, we are conditioned to looking over our shoulders to see who or what is about to destroy it. In other parts of the world the destruction of rainforest is going on unabated. Is it any different here?

When I was in far north Queensland in the late 1960s things did not look good. Logging and clearing were going on at such a pace that I had serious doubts that animals and plants with restricted distributions such as the cassowary, tree-kangaroo, golden bowerbird and certain species of primitive plants would survive. Events took me to other countries but these rainforests remained in my thoughts. If in those days you had asked me what are the chances of stopping the destruction, I would have had to admit it was a wonderful dream but there was no chance of it happening.

But it did happen. Thanks to the vision, courage and determination of many people 900,000 hectares of the Wet Tropics, 630,000 hectares of which is tropical rainforest and just about all there is, was inscribed on the World Heritage list. That was in December 1988. There will be no more logging, no more clearing. It was not achieved easily nor without conflict. But that struggle is now largely over. We and the generations that come after us can enjoy, be stimulated by, be intellectually challenged by, can work towards a greater understanding of all life in these tropical forests without fear of their imminent destruction. No longer will we have to look constantly over our shoulders, though we should always keep a watchful eye out.

On 10 March 1990 agreement was finally reached between the Commonwealth and Queensland governments on a Wet Tropics Management Scheme. It functions through a complicated web of authorities, councils and committees. The central work of formulating plans and overseeing their implementation is done by the Wet Tropics Management Agency based in Cairns. Being involved with these rainforests on so many levels, and also having World Heritage as a neighbour, I am deeply interested in its workings. They will be central to Bulurru's future.

The best way to gain some insight into plans for the rainforest is to talk to Peter Hitchcock, the Management Agency's director.

We meet in his office high up in a tall office block. Through large windows there are views of Cairns city and rainforest covered hills beyond. Peter is in his late forties, quietly spoken and eloquent on the subject of World Heritage rainforest. I sense, as we talk, that he is someone who prefers fair minded and reasonable debate to conflict and confrontation. Yet I have the impression that underneath the reasonableness, kindliness even, there is an unwavering resolve to get the Wet Tropics management right.

I tell Peter about my astonishment and delight at the fact that the tropical rainforests are now protected and ask him what he thinks brought this about.

'At first,' he says, 'I think it was a growing awareness in the Australian community about what was happening to rainforest world wide. From there the focus shifted to the Australian situation and the thought was, "Well, if we try to tell these other countries what to do we should look at what we are doing about our own forests first." I think many people then became rather alarmed at what we were still doing to our remaining very, very small patches of rainforest. I think the push for a more conservative approach to the management of our rainforests grew from that.

'I don't think governments take initiatives on these sorts of things. They get pushed along by public opinion. Public opinion was clearly growing and at the national level it was perceived that rainforest conservation would be a popular move. Of course at the local level these moves were very, very unpopular to begin with. But I think that a general acceptance is growing that World Heritage listing was a brave and wise decision that will actually benefit the community.'

What does the management of the rainforest mean, what does it actually do, I ask. Peter explains:

'The best way to describe it is that it consists of three main programs: heritage conservation, planning, and community relations. The three are linked. Community support is very important. We must develop and encourage it. If you have a totally antagonistic community the rest doesn't matter, because it's going to fail anyway. We're looking to the local community for in a sense they are stake holders. They all have an interest.

'On the ground we have the responsibility of heritage conservation. This takes a variety of forms including research and rehabilitation such as replanting, weed eradication, feral animal control and so on. The planning side of things is the key to it all. It tries to combine demands for use of the area with our conservation responsibilities.

'There will be walking tracks through the rainforest. At the moment we are in the process of mapping all the existing trails. Quite frankly it's a bit of a tangle, with tracks going here, there and everywhere. Some are maintained, others are not. So our attitude is, let's have a look at what we've got before we start building new ones.

'With our research we're making sure that we have clear objectives. Some projects have already been funded. But our future funding will be very much directed toward high priority areas. We're still working that out. There will certainly be an emphasis on endangered species of both plants and animals. We will also be looking at various forms of disturbance such as powerlines and mass tourism.'

The Management Agency has been functioning for only a short time. It

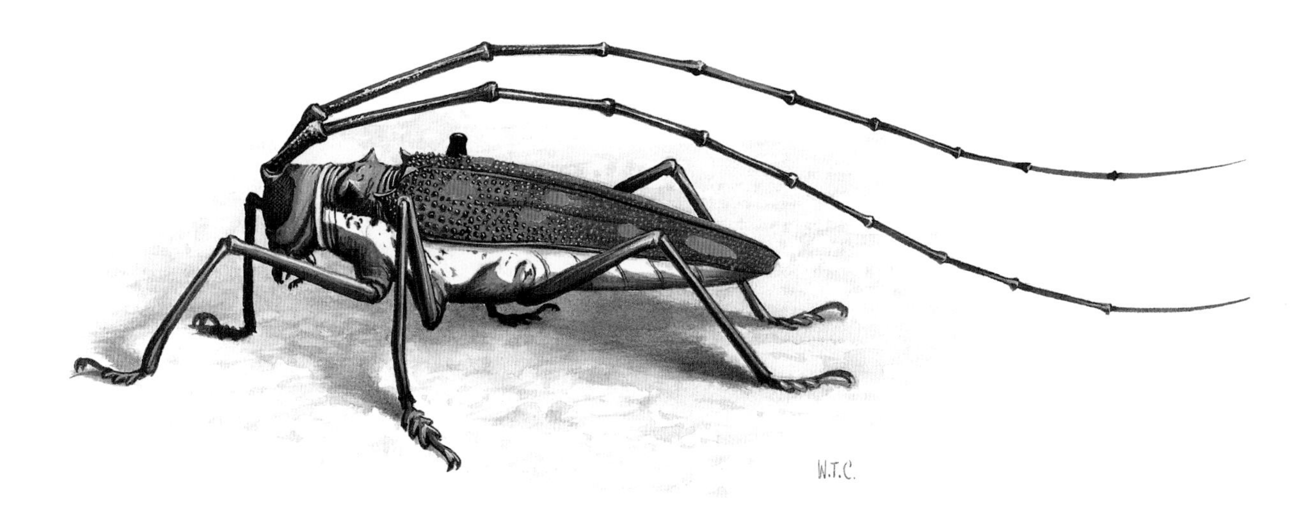

Pink-spotted longicorn beetle.

W.T. Cooper — 92

A lowland rainforest torrent cascades over granite rocks. A water dragon sits on the rocks on the left.

is still preparing plans and working out priorities. So I ask Peter how he envisages the Wet Tropics rainforests in 50 years' time.

'I'd like to think that all the rainforest we have now will still be there. We may even have a little more through rehabilitation and regeneration work. I would like to think that people then looking back over the previous fifty years will say that we have a well managed area, that access is comparatively easy and gives them a full range of experiences of the rainforest. I would like to see a select number of walking tracks that are well planned, well designed and well maintained so that visitors will have quality experiences. I also hope we don't have to wait fifty years to achieve that.'

Visitors both from within Australia and from other countries come to see the tropical rainforest in ever increasing numbers. I ask Peter if he sees this as a problem.

'Yes I do,' he says. 'When it comes right down to it the main threat to the forest will always be people, and they can also kill with kindness. The tourism industry centred around the rainforest is still in its infancy, but it is expanding fast. Many people are now coming from overseas simply to see the rainforest. Regrettably, in some ways, their focus, even before they get here, is largely on the Daintree area. It is so well known internationally that its reputation precedes it. The area is not yet ready for that level of tourism and some degradation is occurring as a result. We are anxious to prevent that and where it does occur to repair it. What we need are a number of innovative approaches to minimise visitor impact. We've made a beginning in this area at Oliver Creek where we built a boardwalk. At the moment there is too great a focus on the Daintree. People are going there through sheer ignorance of other opportunities. There's a lot of other rainforest.'

The World Heritage Wet Tropics is an oddity and a very complex one. Even though it may not contain enormous areas of rainforest if measured against those of the Amazon or southeast Asia, as a single conservation area it is huge and of difficult proportions. It is a narrow strip of land stretching for 450 kilometres along the coast. Its boundary is about 3000 kilometres long. It comprises 620 different parcels of land including 91 freehold properties, 110 leases, as well as state forests and national parks. This is very different from most World Heritage reserves. Most are a single entity or national park like Kakadu or Uluru in the Northern Territory. Planning and management is reasonably straightforward for such places, but here in the Wet Tropics jigsaw it is more difficult. I ask Peter what special problems this poses.

'Well,' he says, 'it means that the actual management of the area is very much in the hands of the agencies—mostly the Forest and National Parks Services but also the Lands Department, the Department of Defence and the Electricity Commission—and the private landholders. We provide the umbrella protection and eventually the umbrella management guidance through specially drawn up plans.'

And how do you see your own role, I ask.

'Not necessarily as the ship's captain but more its navigator charting a course of management for the Wet Tropics,' he replies. 'And rather than being a single ship, I see it more as an armada of different ownerships, all keeping together and pointing in the same direction if we possibly can. It's no easy task.'

As always I am interested in what brings someone to work in or with rainforest, what his or her fascination is. I ask Peter to tell me something of his background.

'I grew up in a little town called Gloucester on the mid-north coast of New South Wales. There are rainforests all around there and it used to be a fascination for me to go into them. They're such a contrast to the surrounding, cleared agricultural land. The rainforest was very special to me then and continues to be special.

'The rainforest was one of the drawcards that brought me up here. The other thing, of course, was the challenge of shaping a large World Heritage reserve into a truly well managed place. This is the third World Heritage area I have been involved with and all of them were to do with rainforest. I worked with Tasmanian rainforest where I was party to putting together the report that resulted in World Heritage listing. Then I did similar work in New South Wales. So I've had exposure to cool temperate, warm temperate and subtropical rainforest. It was a natural progression to move up here, to the tropics. I guess that begs the question: where to from here? I have a great interest in rainforest at a global level. I feel that by setting the example here we can contribute to rainforest conservation in other countries, particularly if we can establish an ecologically viable, well managed tourist industry. We can then demonstrate that there is an economic base for conserving rainforest.'

Peter told me earlier that he had just returned from an international conference on the conservation of the natural environment. I ask him how the rest of the world regards Australia's efforts at conserving rainforest.

'Interestingly enough,' he says, 'at the international level most people would not be aware that Australia has any rainforest at all. But in conservation circles they are very much aware of it, especially in the Wet Tropics. This international interest, as far as I know, is largely derived from the controversial aspects of the World Heritage nomination, which was contested by the then State government. As a result, a lot of people asked me how it is working. In the international conservation community the Wet Tropics are seen very much as a potentially important world model. They are naturally concerned that if we can't get it right there isn't much hope for a lot of other countries. So I feel a certain responsibility to get it right. I'm confident that we can.'

In the end I come back to the question that brought me here and that is: how secure is the tropical rainforest? Will it become a political football, being put in and out of World Heritage as governments change? Peter thinks not.

'I've seen these sorts of threats in other places,' he says. 'But when push comes to shove nothing happens. Governments realise that the community out there is going to create such a stir if something happens to the rainforest, that they would back away. Also I think one of the benefits of World

Heritage listing is that if there is any significant threat to it, there is this worldwide network that goes into action. It can bring diplomatic pressure to bear which should not be underestimated. So in many ways World Heritage listing does provide effective protection, not because it has any legal teeth but because the international community is watching. Other countries are expecting good things, big things, out of the Wet Tropics. They don't want us to fail. They want us to be an absolute model of success.'

Flying-foxes

Bulurru, 11–12 December

The flying-foxes' daytime roost, or camp as it is usually called, is in a gently undulating area of forest in the northeast corner of Bulurru. A few emergent trees project above the canopy and a small stream meanders through it. During the cooler months, between May and September, only a few of the giant bats were in residence. But since mid-September there has been a marked build-up in numbers. Increasingly I can hear the noise of their squabbles and if the breeze is from the right direction I get a faint whiff of their distinctive odour. In the evenings I see them stream out of their camp in search of flowering and fruiting trees in the surrounding forests. By now there must be many tens of thousands of the fruit-bats. A count of any accuracy is almost impossible, spread out as they are over several hectares of dense forest. When they leave at dusk they stream out in different directions and continue well after it is too dark to see them.

In late September the first young were born; tiny perfect replicas of their parents complete with voluminous wings. But at birth they were unable to fly. For the first month of their lives they had to cling tightly to their mothers. The babies have especially sharp, recurved milk teeth, later shed, which enable them to fasten onto their mother's teat, of which there is one under each wing. Extra-sharp claws on their feet give the young a firm grip on the females' fur. Hanging on with teeth and claws to their mothers' underside, the young are carried everywhere, even on a 30- or 40-kilometre flight to the adults' feeding places. The season of birth continued for a few months and is now all but finished.

At mid-morning, when it is already hot, I approach the camp. For some time I have noticed the noise and the odour which, while getting quite strong now, could not be said to be either overpowering or offensive. The first bats I see hanging upside down in low shady trees are old males. As they hear my footsteps they look at me over their folded wings through half-closed eyes. They utter a few subdued but high-pitched notes: their call to say there is possible danger. These peripheral males are the camp's early warning system. As soon as they raise the alarm this corner of the camp falls silent; all the bats are watchful. After less than a minute the guards wrap their wings more tightly around themselves and go back to sleep. Quarrelling screeches and screams break out again.

Slowly I walk into the main camp. My presence causes little concern. Earlier in the year the bats would take off as soon as I set foot among them. The taller trees are packed with rows and rows of black bodies. The bats' comings and goings have stripped the outer branches of leaves and it is quite light and open inside the camp.

All the bats in this colony are of the same species, the spectacled flying-fox or fruit-bat, so named because of the rings of pale fur around their eyes, their 'spectacles'. They look at me through large round dark brown eyes, their sensitive long muzzles sniff in my direction and their expressive triangular black ears are constantly twitching. Their bright alertness has an air of intelligence about it, but without the knowing calculatedness that characterises such animals as rats and crows. I am given only a brief inspection and soon the bats resume their normal preoccupations. Most go back to sleep, suspended from just one foot, the other foot and their heads tucked into the folded wings. Most females have young. The older infants, which can already fly, hang beside their mothers. Some of the smaller babies suckle. Here and there females hold their newborn wrapped in the embrace of their wings. One such tiny young is very boisterous and cannot be contained. It exercises its wings with great energy and beats its long-suffering parent about the head. She eventually stops the youngster by licking its wings.

All through the camp bats groom themselves, scratching with one foot

Spectacled flying-foxes leaving their camp at dusk.

and licking themselves all over. They rub their wings over oily glands in the pale blonde fur on their shoulders to keep these membranes flexible and waterproof. It is hot out in the sun yet many of the flying-foxes hang right out in the open. They fan themselves with their wings, but even so their black bodies, especially their large hairless wings must get very hot. All the grooming, fanning and wing flapping gives the camp an energetic, animated look. There is also discord. I am startled by a sudden piercing, gurgling shriek and the sound of scuffling right above me. A well-orchestrated 'fight' has broken out with lots of sparring and snapping with hooked claws and sharp teeth. But the combatants never come to actual blows.

It is the time of year for the males to return to the centre of the camp and set up territories among the females. During the season of birth males and females were segregated. Now, family groups are formed, though mating will not occur for some months. The squabbles are mostly between males disputing territorial ownership on a length of tree branch. But females will also 'attack' males if they come too close or become too amorous. One of the quarrelling males above me takes off, only to land elsewhere and start another chain reaction of mock battles and screams. I leave them to their noisy squabbles.

At dusk I return to the camp. I position myself under some small trees at Bulurru's eastern boundary. It is a sharp divide; my place is rainforest, my neighbour's grassland. The sky is an aquamarine colour invaded by orange flares from the setting sun. To the north a storm lingers; an ink-black cloud lit from within by flashes of lightning. To the east, beyond my neighbour's paddocks, there is more rainforest leading up to Mount Bartle Frere's purple bulk. The flying-foxes in their camp about 50 metres away are noisy.

Twilight. The flying-foxes no longer shriek but only murmur. This too dies down. A few moments later a single black shape flies out from the rainforest. It quickly circles back and is met by a dozen more. Most fly off strongly, some turn back as if reluctant to fly over cleared land. Within seconds a steadily flying broad column of fruit-bats streams out of the camp, flying low directly over me. The sound of their beating wings makes a subdued humming noise. As always I am impressed by the bats' size. Their wingspan is about 1.2 metres. They fly expertly, beating their wings at about two strokes per second, and reaching speeds of 35 kilometres per hour. Apart from air being pushed by countless wings and an occasional soft-voiced communication, the stream of thousands upon thousands of bats is remarkably quiet. As they fly their ears are pressed back against their heads, but their long muzzles still have a fox-like look about them. They do indeed look like flying-foxes, as long as you do not consider their tails. The bats do not have any.

A wide, unbroken column of bats now stretches from the colony to the lower slopes of Bartle Frere, three or four kilometres away. Flying-foxes navigate entirely by sight. When they reach their feeding grounds, as much as 40 or 50 kilometres away, it will be dark. The bats, however, have excellent night vision and can find their way on the darkest nights. They find their food trees, with nectar-rich flowers or with ripe fruit, by scent or by the sound of the voices of other flying-foxes already feeding.

Darkness falls rapidly. Stars appear. Still the columns of bats pour out. About half an hour after the first bat left the camp, the exodus slows to a trickle. I can hear the kookaburras that roost in a tree beside the house give their last laugh for the day. There is still a little light in the sky and I notice a lot of bats flying low over the canopy, even between the trees. They seem to me to be somewhat smaller than the ones that had been streaming out. Three of them circle around me, at about head height, only a few metres away. They are very curious. They fly on a little and land in a small pink ash on the forest edge. Others join them till there are 40 or 50 tightly packed along the branches. All daylight has gone now and I can hear rather than see these bats. I shine my torch on them. The tree lights up with the shine from scores of eyes moving and jostling. In ones and twos the bats fly off until only about half a dozen are left. These are indeed very small, very young bats, clinging awkwardly to the branches as they look into the light. These were born earlier in the season and are now too old to be still carried by their mothers but too young to fly out to the feeding grounds. I had not realised that bands of juveniles roam around the camp at night. The tens of thousands of females and their young, which still depend on their mothers for nourishment, will probably return to the same tree in the camp and then locate each other by sound. Incredibly, each female can recognise the cries of its own offspring in the camp's shrill and constant cacophony. Only when they are over three months old will the juveniles leave their mothers. They will then form dense packs in their own corner of the camp.

When I get back to the house there is a busy tumult of fruit-bats in the rusty fig. Broken-off leaves rain down and small branches snap under the weight of too many bats, as the fig is stripped of its fruit. Many of the raiders are small young, perhaps on their first flight.

This morning I am up a couple of hours before sunrise. Already the camp is noisy with the sounds of many thousands of bats back from the night's foraging. The rusty fig is empty. A crescent moon hangs low over Bartle Frere. The stars are bright. A boobook owl calls in a distant valley. I return to the place where I watched the bats leaving last night. Once again I am inspected by a bank of inquisitive youngsters. I can just make out the silhouettes of mature bats flying back to their camp. When I get nearer, the full force of the sound hits me: an indescribable, continuous high-pitched shrieking with an occasional nearer or louder scream rising above the general dissonance. Surely most of the bats have already returned?

A faint red glow appears to the south of Bartle Frere, putting a blush on the band of mist garlanding the mountain. The stream of returning flying-foxes is now clearly visible. Most fly low, steadily, following the creeks. They come in twos and threes, then suddenly there are knots of 20 or more followed by bands of several hundred. Some arrive back flying high then glide lazily down or dive in a steep, fast zigzag that makes their wings vibrate like sails flapping loose in the wind.

A male purple-crowned pigeon tests a hazel-wood fruit for ripeness before swallowing it.

Inside the camp there is much noisy scuffling and shoving along the branches. Most bats, however, are asleep despite the noise. The females have found their offspring, many of which are suckling. Others are being groomed. The shrieking comes from new arrivals trying to find a roosting place over the objections of those already in residence. The kookaburras at the house wake and call. Channel-billed cuckoos trumpet in the distance. The sun rises.

Return of the Pigeons

BULURRU, 15 DECEMBER

The masses of fruit on the white hazelwoods close to the verandah are ripening. They are a little earlier this year and I am hoping for another day of wompoos. It may not be quite as dramatic as last year as the flying-foxes have finished off the rusty figs. Fortunately the bats have no taste for the hazelwood's purple berries. The goyas failed to fruit this year but are decked out in glossy yellow-green new foliage.

Late in the afternoon rain falls gently, freshening but not drenching the trees and the grass. When the sun's rays are low and its light yellow the sky clears. The foliage sparkles. Suddenly four white-headed pigeons land in the topmost branches of a small hazelwood. They sit bolt upright, their whiteness contrasting with the dark green hill beyond. More and more pigeons tumble out of the sky. Some land in a flurry of wings in the fruiting tree. Others overshoot as they rush in; there must be more than 20 milling around. At least seven wompoos join in the fray. Two pairs of purple-crowned pigeons, so much smaller, but no less colourful, land in the lower branches. Wompoos, flashing the bright yellow undersides of their wings, try to chase the others off, but to no avail. Many more land in the trees only a few metres from the verandah. The birds feed in a hurry. Each fruit is tested for ripeness by gently squeezing it in the bill before being pulled off and swallowed. Some of the birds overbalance as they reach for the fruit on the thin outermost twigs and flail their wings to keep their balance. As they thrash about they send showers of droplets in all directions. The birds are so close that I can see the white-heads' red eye-rings and coral-red feet. Their backs are not the uniform dark grey they appear from a distance but shine with iridescence, sometimes green, sometimes purple depending on the direction of the light. A flock of king parrots lands in a tobacco bush behind the pigeons, adding their brilliant red to the rainbow colours of the fruit-eaters.

Having filled their crops the pigeons retire to the tall quandong to preen and to doze. Once again, word has got around that the fruit is ripening. The year has come full circle.

It is the same as last year, yet it is different. It is the same in that

Male spangled drongos singing and posturing at each other.

pigeons are eating the fruit, that the quandong is their lookout, that the grey goshawk uses it for an overview of his domain. It is different in that the white-heads now dominate, that the flying-foxes monopolised the fig, that the fruit is earlier, and that the topmost branches of the quandong have died and will probably blow down in a storm before long. As always in nature there is dynamic change within eternal cycles of death and renewal.

I must now ask myself, did I come to terms with the tropical rainforest during the course of the year? Does the mass of green out there make some kind of sense now? Through observing the passing of the seasons and becoming more familiar with the plants and their patterns of growth, in other words by being drawn into the elemental forces at work here, my understanding and appreciation of the forest have deepened. The animal life, the pigeons, flycatchers, robins and honeyeaters around the house, the tooth-bills and golden bowerbirds in the forest, the green possums in the rusty fig, the bandicoots and pademelons tearing up the grass at night, the python that lives near the creek, they are no longer 'specimens' to be dispassionately observed. They have become individuals, neighbours, and part of my everyday life. The place does make some kind of sense. I have a general idea of how its forests came to be here, about their great age and how they fit into the vegetation pattern of Australia as a whole. This ancientness and the Australianness, concentrated in this tiny wet corner of the huge continent, have for me been the great discoveries of the last year. It has been a journey into Gondwana.

It is dusk as I write this. Grey-headed robins call in their penetrating piping voices and skip over the lawns after insects as they do every evening. I can hear the song of spangled drongos. Two of the birds face each other in duet. One has his crown feathers flattened but those on the side of his face are raised, giving his head a peculiar dish-shaped appearance. A third drongo sits a little way away; possibly a female watching the males in a duel of song over her favours. Every day there are new things to see, new secrets to discover.

For years the identity of this vine was a mystery to me. The fruit was recently identified as belonging to Salacia disepala. *As a bonus I discovered that the seed is the food of the caterpillar of the Australian plane butterfly.*

Plant Names

alder, blush .... *Sloanea australis*
  buff .... *Apodytes brachystylis*
  rose .... *Caldcluvia australiensis*
almond, red .... *Alphitonia whitei*
ardisia, mountain .... *Ardisia brevipedata*
aroid, climbing .... *Rhaphidophora* sp.
ash, pink .... *Alphitonia petriei*
aspen, white .... *Acronychia vestita*
banana, native .... *Musa* sp.
banyan, Australian .... *Ficus virens*
bark, almond .... *Prunus turneriana*
basswood .... *Polyscias murrayi*
  ivory .... *Polyscias australiana*
bean, bitter .... *Pouteria castanosperma*
  black .... *Castanospermum australe*
  burny .... *Mucuna gigantea*
  matchbox .... *Entada phaseoloides*
  scarlet .... *Archidendron ramiflorum*
beech, buff .... *Irvingbaileya australis*
  northern white .... *Gmelina fasciculiflora*
  red .... *Dillenia alata*
bells, misty .... *Agapetes meiniana*
berry, crimson .... *Guettardella tenuiflora*
  lolly .... *Salacia chinensis*
bleeding heart .... *Omalanthus novo-guineensis*
bollywood, brown .... *Litsea leefeana*
  white .... *Neolitsea dealbata*
bolwarra .... *Eupomatia laurina*
box, kanuka .... *Tristaniopsis exiliflora*
bush, bridal .... *Breynia stipitata*
  fart .... *Breynia stipitata*
  spice .... *Triunia erythrocarpa*
candlenut .... *Aleurites moluccana*
carabeen, grey .... *Sloanea macbrydei*
  red .... *Geissois biagiana*
  rusty .... *Aceratium ferrugineum*
cedar, red .... *Toona ciliata*
celerywood .... *Polyscias elegans*
cherry, river .... *Syzygium tierneyanum*
  Herbert River .... *Antidesma bunius*
coffee, native .... *Breynia stipitata*
coralwood .... *Adenanthera pavonina*
corkwood .... *Melicope elleryana*
cunjevoi .... *Alocasia macrorrhiza*
delarbrea .... *Delarbrea michieana*
Dutchman's pipe .... *Aristolochia* sp.
ebony heart .... *Elaeocarpus bancroftii*
evodia .... *Melicope* sp.
fern, basket .... *Drynaria* sp.
  bird's nest .... *Asplenium nidus*

common tree .... *Cyathea cooperi*
elkhorn .... *Platycerium bifurcatum*
king .... *Angiopteris evecta*
scaly tree .... *Cyathea cooperi*
fig, banana .... *Ficus pleurocarpa*
cluster .... *Ficus racemosa*
curtain .... *Ficus virens*
deciduous .... *Ficus virens*
rusty .... *Ficus destruens*
white .... *Ficus virens*
gardenia, hairy .... *Randia hirta*
mountain .... *Gardenia merikin*
ginger, common .... *Alpinia coerulea*
goya .... *Guioa lasioneura*
gum, flooded .... *Eucalyptus grandis*
hazelwood, white .... *Symplocos cochinchinensis*
ivory wood .... *Siphonodon membranaceum*
kurrajong, tulip .... *Franciscodendron laurifolium*
lantana .... *Lantana camara*
lasiandra, native .... *Melastoma affine*
laurel, Murray's .... *Cryptocarya murrayi*
rusty .... *Cryptocarya mackinnoniana*
leaf, beauty .... *Calophyllum inophyllum*
lily-pilly, powderpuff .... *Syzygium wilsonii*
mahogany, brush .... *Geissois biagiana*
mangosteen, mountain .... *Garcinia gibbsiae*
maple, Queensland .... *Flindersia brayleyana*
milkwood, grey .... *Cerbera manghas*
myrtle, apricot .... *Pilidiostigma tropicum*
nun, blue .... *Delarbrea michieana*
nutmeg .... *Myristica insipida*
oak, Atherton .... *Athertonia diversifolia*
ochna, native .... *Brackenridgea nitida*
ochrosia .... *Neisosperma poweri*
orchid, buttercup .... *Dendrobium agrostophyllum*
oak .... *Dendrobium jonesii*
pencil .... *Dendrobium teretifolium*
spider .... *Dendrobium tetragonum*
palm, Alexandra .... *Archontophoenix alexandrae*
Atherton .... *Laccospadix australasica*
black .... *Normanbya normanbyi*
fan .... *Licuala ramsayi*
orania .... *Oraniopsis appendiculata*
walking stick .... *Linospadix* sp.
palm-lily .... *Cordyline* sp.
pandan, climbing .... *Freycinetia* sp.
panax, climbing .... *Cephalaralia cephalobotrys*
pepper wood .... *Cinnamomum laubatii*
pine, brown .... *Podocarpus dispermus*
China .... *Goniothalamus australis*
bull kauri .... *Agathis microstachya*
plum, cassowary .... *Cerbera floribunda*
Davidson's .... *Davidsonia pruriens*
poison .... *Pouteria castanosperma*

pothos .... *Pothos longipes*
quandong, blue .... *Elaeocarpus angustifolius*
  buff .... *Peripentadenia mearsii*
  Kuranda .... *Elaeocarpus johnsonii*
  tropical .... *Elaeocarpus largiflorens*
rattle, red .... *Triunia erythrocarpa*
sarsaparilla, Austral .... *Smilax australis*
sassafrass, scentless .... *Daphnandra repandula*
satinash, bumpy .... *Syzygium cormiflorum*
  grey .... *Syzygium gustavioides*
  paperbark .... *Syzygium papyraceum*
  plum .... *Syzygium wilsonii* spp. *cryptophlebium*
  roly-poly .... *Syzygium endophloium*
  scarlet .... *Syzygium erythrocalyx*
  weeping .... *Syzygium apodophyllum*
silky oak, blush .... *Opisthiolepis heterophylla*
  briar .... *Musgravea heterophylla*
  brown .... *Darlingia darlingiana*
  crater .... *Musgravea stenostachya*
  northern .... *Cardwellia sublimis*
  pink .... *Oreocallis wickhamii*
  rose .... *Placospermum coriaceum*
  Whelan's .... *Macadamia whelanii*
silkwood .... *Flindersia* sp.
stinger, Gympie .... *Dendrocnide moroides*
supplejack, white .... *Ripogonum album*
tamarind, Boonjie .... *Diploglottis bracteata*
  native .... *Diploglottis diphyllostegia*
  pink .... *Toechima erythrocarpum*
  Topaz .... *Synima macrophylla*
tarwood .... *Semecarpus australiensis*
touriga, blush .... *Calophyllum sil*
  satin .... *Calophyllum inophyllum*
tree, flame .... *Brachychiton acerifolius*
  umbrella .... *Schefflera actinophylla*
tuckeroo, brown .... *Cupaniopsis flagelliformis*
tulip oak .... *Argyrodendron* sp.
umbrella, blue .... *Mackinlaya macrosciadia*
vine, bellbird .... *Melodinus australis*
  fan flower .... *Scaevola enantophylla*
  jungle .... *Neosepicaea jucunda*
  lacewing .... *Adenia heterophylla*
  October glory .... *Faradaya splendida*
  pepper .... *Piper* sp.
  zigzag .... *Melodorum leichhardtii*
wait-a-while, common .... *Calamus australis*
  fishtail .... *Calamus caryotoides*
  yellow .... *Calamus moti*
waratah, tree .... *Oreocallis wickhamii*
walnut, brown .... *Beilschmiedia tooram*
  poison .... *Cryptocarya pleurosperma*
  Sankey's .... *Endiandra sankeyana*
wilkiea .... *Wilkiea* sp.

Note: The common names of trees may apply to several different species.

Index

Note: Page numbers in italics refer to illustrations.